# Development is the Name for Peace

*Selection of the Proceedings
1988 Conferences of the
Schiller Institute*

**SCHILLER
INSTITUTE**

*Washington, D.C.*
*1988*

# Development
# is the Name for
# Peace

## *Selection of the Proceedings 1988 Conferences of the Schiller Institute*

January 30-31, Andover, Massachusetts

March 26-27, Cologne, Federal Republic of Germany

Other titles by the Schiller Institute:
*The Hitler Book*, edited by Helga Zepp-LaRouche (1984)
*Rescue the Western Alliance!* (1984)
*Colonize Space: Open the Age of Reason* (1985)
*Friedrich Schiller, Poet of Freedom* (1985)

On the cover: Construction of a hydroelectric dam in Sierra Leone.
Magnum Photo/Gaumy

Cover design: Rosemary Moak
Book design: Virginia Baier
Compositor: World Composition Services
Printer: PMR Printing Company

Schiller Institute
P.O. Box 66082
Washington, D.C. 20035-6082

ISBN: 0-9621095-0-9

# TABLE OF CONTENTS

# Contents

HELGA ZEPP-LAROUCHE

# For a Just New World Economic Order! Development Is the Name for Peace

The most essential premise on which the Schiller Institute was built is that the Western Alliance will survive only if the current bankrupt and immorally decaying world monetary system is replaced promptly by a just, New World Economic Order. On October 19, 1987, "Black Monday" confirmed the prediction of a worldwide financial crash by American presidential candidate and economist Lyndon LaRouche, who had forecast this crash in May, precisely for the following October.

Yet, though the desolate condition of the world economy could leave no doubt that this was only the first phase in the potentially greatest collapse of financial markets of all time, the heads of the Western industrial nations and the representatives of the leading financial institutions spoke only of a "correction" of the markets.

The leadership of the nations belonging to the Organization of Economic Cooperation and Development did nothing to protect industry and agriculture from the consequences of the crash, not to speak of the protective measures, which might have prevented the broader, and still more dramatic phases

Helga Zepp-LaRouche *is the chairman of the Schiller Institute Executive Board.*

of the collapse. Instead, the most brutal austerity programs were implemented, which, in the face of the world hunger catastrophe, have only led to the insane shutting down of industry and agriculture and intensified the desperate distress in the developing nations.

The world economy was again in the first period of a depression, as the developing sector was now even more harshly dazed—and those responsible in the leadership had evidently learned nothing from history. To the contrary, the austerity policy of Hitler's Finance Minister Hjalmar Schacht was even explicitly cited as the solution.

The reason for this clinging to an obviously incompetent economic and financial policy, and to the deliberately circulated illusion that the markets had again calmed themselves, lay not exactly in the stupidity of economic policy *per se*. It lay much more in the immovable determination of the leading banking and financial circles to hang onto their power at any cost, even if this should mean the end of Western civilization.

The American Establishment had resolved that George Bush should become the candidate of the banks and the next President of the United States. Therefore, the inevitable total collapse of the financial system must be postponed, at any cost, until after the U.S. election in November. Accordingly, not only would maximum pressure be brought to bear on Japan and Western Europe, to have their central banks bail out the dollar over this time period, but it was resolved, through the glorious summit between Reagan and Gorbachov, to divert attention from the reality of the depression in the United States, and thus to secure the election for Vice President Bush in the wake of the "Peace President," Ronald Reagan.

Thus, on Pearl Harbor Day in the United States, December 7, 1988, the INF Treaty was signed between Reagan and Gorbachov in Washington, D.C. This treaty signified on behalf of the United States nothing other than the intent to withdraw from Western Europe and to give it over to Moscow's hegemony. This means, however, that the U.S. Establishment is prepared to give Moscow world domination in a New Yalta agreement—with the Finlandization of Western Europe as a consequence—for the sole purpose of hanging onto a bankrupt banking system.

When the Schiller Institute was founded four years ago, we warned that the sharpening world financial and economic crisis threatened to lead to the decoupling of Western Europe from America, and at that time questioned many on this forecast. Today, unfortunately, there remains no question that we were right with our initial premise; namely, that the failure to face the necessity to overcome the depression through a just, New World Economic Order, can lead to the downfall of the Western Alliance and, with it, all that has made up Western civilization of the last 2,500 years.

If this reform is not accomplished very soon, human society, in fact, stands before a many-layered hell. Already, the entire African continent is being officially written off by the leading banking circles. That continent is threatened with extinction, if a large-scale Marshall Plan is not immediately implemented, which effectively confronts the combination of AIDS, hunger, and locust plague.

All of Latin America threatens to be transformed into an uncontrollable witch's cauldron, in which the private armies and guerrillas of the drug mafia have a free hand; indeed, these gangs still receive support from the U.S. administration, because the preservation of the world financial system depends on drug money. Likewise, an ever larger part of the youth is sacrified to the drug epidemic, in the lands where drugs are cultivated, as well as in the so-called industrial nations.

Without these reforms, the financial means for the research of a real cure for AIDS will be just so much smaller, as will be the money for caring for the swelling number of victims of this potentially mankind-threatening epidemic. The unconcealed propagandizing for euthanasia gives a foretaste of what could follow, if the health insurance fund is empty and the pensions are no longer paid. The spread of satanism and occultism, which has not stopped even at the level of school children, belongs just as much to this picture, as the escalation of terrorism, which, indeed, readily finds a fertile ground in the lack of development. No, there remains unfortunately no question, that the future will be a good approximation of hell, if the depression intensifies. Up-and-coming dictators, be they successors of Gorbachov or the representatives of the supernational institutions, will have nothing to lose.

But why should man permit that mankind be destroyed, because of the policies of a relative few, whose power-madness blinds them to the consequences of their policies, when the absolute majority of mankind is urgently longing for development and a future worthy of mankind? Therefore, the Schiller Institute called world leaders into action.

Leading figures from North and South America, Asia, Africa, and Western Europe took part in two conferences, sponsored by the Schiller Institute, on the theme of the New World Economic Order. The first conference was convened in Andover, Massachusetts, January 30-31, 1988, not far from the historic Bretton Woods, to discuss the parameters of a new "Bretton Woods System," which, this time should be without the built-in injustices of the old monetary system. At this conference, it was demonstrated, how an agreement is possible between so many different nations, with relative ease, when an economic program facilitates the development of all, North and South. It became evident, why all the many United Nations conferences have fallen apart in disunity—precisely because this programmatic basis was lacking. In Andover, the participants spoke out of passionate responsibility, to organize an ever-wider circle worldwide, in the coming weeks and months, for a New World Economic Order.

So followed a broader conference in Cologne, West Germany, on March 26 and 27, 1988. This time, among other issues, the catastrophic economic situation in the East bloc was elucidated. Further, the urgent necessity of bringing the economic policy of the West into harmony with the principles of the Christian ethic was discussed.

Mankind has reached the point at which our moral ability to survive will be tested. If what Pope John Paul II identified in his recent encyclical as the "Structures of Sin" should win out, if egoism and the power-hungry should be victorious, then our Western civilization, for a variety of reasons, will go under. It will collapse, as did the Roman Empire, whose so-

# Introduction

called "elite" were also too decadent to notice the dangers, internal and external, and to deal with them.

The Schiller Institute will do everything in its power, to overcome this existential crisis, through a cultural and moral renaissance, and will establish a just, New World Economic Order, which will make possible the dignity of all men on this planet.

# GREETINGS

The following greetings are some of the messages that were received and read at the Schiller Institute's January 30-31, 1988 conference on the New World Economic Order.

## His Excellency Dr. K. D. Kaunda

*President of the Republic of Zambia, Chairman of the Organization of African Unity*
Your kind invitation has been received. I regret the delay but this was due to the fact that I thought I could make some readjustment in my overcrowded programme. However, this is not possible and I just have to regret—very much regret—that I will be unable to join you on this very important occasion at which you will discuss matters affecting the new world economic order, the way out of the depression. I am awfully sorry about this. I wish you every success.

## His Excellency Abdou Diouf

*President of the Republic of Senegal, Dakar*
. . . In 1979-1980 . . . the world economy underwent a profound commotion which plunged it into grave difficulties which, despite the attempted solutions which were unleashed at that time, has not yet been dispelled. One can even say that the situation has deteriorated.

# Greetings

. . . Today, more than ever, an imperious duty is imposed on all those who make up the international community, which must redouble efforts, in reflection and action, so as to overcome the obstacles and bring about the economic reform in which both all the industrialized nations and the developing countries are engaged.

Your work is going to be the occasion for pursuing reflection on the problem of reforming the world economy and in particular, on the thorny question of the Third World debt.

There can be no more important questions than these, and I must solemnly thank the organizers of this meeting for having launched this initiative and for having desired that I associate myself with it.

I also salute the participants in this conference for their mobilization around essential questions for the future of the world economy.

The grave problems inherent in the debt burden have heavily mortgaged the development efforts of the debtor countries, and a good number of these countries are on the verge of bankruptcy. . . .

According to the data of the World Bank, trends in the area of resources indicate that gross capital inputs over the long term in the direction of the developing countries have substantially gone down between 1980 and 1985. On the other hand, the prices of basic products have sunk, the which has worsened the already serious adjustment problems which these countries were confronted with. The African economy, which has suffered various shocks of the crisis, needs to be reformed so that the irreparable is not produced.

But it is necessary to recognize that African countries, delivered over to their own resources and to the fluctuations of the external environment, have very little chance for getting out of this crisis because of the weakness of their economic structures. . . .

But as a whole, the concrete commitment of the international community on Africa's side appears timid, in particular, the debt questions remains unsolved and continues to be posed in all its gravity. . . .

Ladies and Gentlemen, through this message, I have wished

to underline all the interest which is attached to the holding of this conference. I wish for full success in your work.

## Monsignor Mario Pimpo

*Prelate of Honor of His Holiness, Vicarage of Rome*
I invoke from God heavenly blessings for your commitment to build a better world, through a new economic order, founded wholly on Christian principles. I wish the comforts of success to Mr. LaRouche's efforts to safeguard the Good and progress of his fatherland.

## Senator Vincenzo Carollo

*Former member, Federal Parliament of Italy*
. . . I express my full support for the Schiller Institute conference initiative which aims at justice and good, the basis for true peace in the world. As long as the rich try to prevail over the poor, justice will be mangled and peace will be in servitude of the weak towards powerful men and peoples. This would not be peace, but permanent war of wealth against poverty. *Populorum Progressio* asserts: "The economy is in the service of man." . . . Let this healthy teaching become the basis for international relations and then peace will be a good for all, based on justice and the reciprocal respect of peoples.

## Honorable Anthony Beaumont-Dark

*Member, British Parliament*
This conference . . . could not have come at a better time and be more necessary, before the economic situation gets so bad as not to be able to respond to positive action instead of panic.

The two big problems we are facing are, first, the U.S. deficit and secondly, the Third World debt. Unless an answer be found, there will undoubtedly be a world financial crisis of frightening proportions. Money is being created without the backing of real wealth. That is what causes inflation, and, in the end, slumps.

The greatest problem that faces the international banks is the several hundred billion dollars of Third World debt, that is not only unlikely to be able to be repaid, but even unlikely for the interest to be repaid. And unless this problem is tackled, many so-called solid banks are bluntly doomed because they are built upon sand.

The current U.S. deficit is not just the U.S.'s problem, it has to be looked upon as a problem that vitally concerns Japan, Germany and the United Kingdom. Unless we all play our part, the United States may go in for trade protectionism, which could ruin any chances of finding a logical and orderly way out of the problem we all face.

Fiddling with exchange rates, which seems almost to be the only "answer" that central bankers come up with, is not an alternative for real answers that have to be found, and can be found, if leaders of the Free World will actually think about the problems more than just their own political ground.

With my best salute to the conference and its participants.

Steven J. Lewis

*Stockbroker, London Stock Exchange*

. . . From the point of view of the central banks, the cheap dollar, combined with the associated fall in the U.S. stock market, which is a consequence of the investors' perception that nothing was going to be done to help the U.S. currency, affords foreign interests an opportunity to buy up remaining productive U.S. assets at low costs.

The uncontrolled fall of the dollar and in the stock market, may threaten a prolonged depression, in which the political outcome, especially in the less developed countries would be uncertain. This is why the central banks, even the previously hard-line Bundesbank, joined in actions after the October 19 Black Monday to flood the money markets with liquidity. A measure of stability has been restored to the financial markets which gives a platform from which foreign purchases of United States assets can now proceed.

Central bank policy is likely to be opposed, however, to any significant recovery in Wall Street which would increase the cost of U.S. assets, and it is also opposed to any rally in the dollar, which would have the same effect. Consequently, should the U.S. dollar show any sign of revival, the Group of Seven central banks would probably agree on actions to be taken to cap the U.S. currency.

From the U.S. bank standpoint, the rise and fall of the dollar has been very acceptable. In the period of dollar strength, the banks benefited from their lending to the expanding service sector. The dollar attracted international short-term funds which provided the basis for a massive expansion in the balance sheets of the bank. Since 1985, the banks have benefitted from the weak dollar, because this has lessened the pressure on their debtors in the less developed countries to renege and default on their borrowings.

The dollar is now so weak, however, and the stock market fall has so revealed the fragility of the financial structure that at least some of the U.S. banks are struggling to maintain their world market share and profitability. There is in this situation a strong likelihood that the community of interest between the central banks on the one hand and the U.S. commercial banking interests on the other, will break down. And in that event, the gyrations of financial values will be inimical to solving the basic economic problems.

The basic economic problem is that the industrial sector is eroding through the lack of adequate investment in productive assets. Rapidly fluctuating financial markets will not fulfill the function of allocating capital in a rational way to productive investment. Reform of the world monetary system and of the financial structure is, therefore, a precondition for the resumption of industrial growth.

## Senator Jamil Haddad

*National president, Brazilian Socialist Party; member, Foreign Debt Commission of the Brazilian Federal Assembly*
. . . It is impossible for Third World nations to continue being exploited and looted through the foreign debt mechanism, when the creditor institutions know that floating-interest-rate loans would never be paid.

This is why I praise your efforts to modify the world economic order and build a just one, under the principles laid down by Lyndon H. LaRouche in *Operation Juarez,* which, upon analysis, I agree with. Thus, I would have liked very much to follow closely the proceedings of this conference of great social importance for today's world.

## Waldo Lima

*Secretary general of the Nationalist Parliamentary Front, PMDB, ruling party of Brazil*
I have the satisfaction of knowing that Brazil is represented [at your conference] through the competency and patriotism of the eminent professor, Garcia Munhoz, of the University of Brasilia.

His ideas regarding these very serious problems which grieve most of all the masses of the Third World, which threaten peace, coincide with the ideas about these problems which we have defended in the Nationalist Parliamentary Front, and have been upheld by the countries of the Group of 77 within the United Nations Organization, led by Yugoslavia and the courageous Peruvian President Alan García. Best wishes for a complete success.

## Honorable Virgildasio de Senna

*Federal assemblyman, National Constitutional Assembly of Brazil*
As the Schiller Institute realizes a meeting with the purpose of establishing a new and durable basis for a new economic order, I wish to join this effort. While I am unable to attend, I wish all the participants the greatest success, and I embrace and congratulate each and every one of you for your efforts toward establishing a social and economic order with more justice for all.

## Honorable Sandra Cavalcantti

*Federal assemblywoman, National Constitutional Assembly of Brazil*
. . . I follow your work with keen interest, knowing that it may bring to life the ideals of the unforgettable Paul VI in his

*Populorum Progressio* encyclical, which is the doctrine closest to our political life and social doctrine.

## K.J. McDonald

*State representative (R), Minnesota*
I send my greetings as a fellow supporter of the movement to defend the Judeo-Christian values of the West against those who would wish to depopulate the world through disease and famine, or permit us to fall prey to the Soviets by weakening our commitment to these values.

As a member of the Agriculture Committee of the Minnesota State House of Representatives, I am aware of the potential of our farmers to feed the world. I endorse the effort to do so, as represented by the commitment of the people at this conference, so that, indeed, we may have a new, just, and more humane economic order.

## Ron Thelin

*International vice president of the Cement Masons, AFL-CIO, U.S. and Canada*
Today, our labor force is idle and out of work. Our labor force is living in streets, homeless; standing in food kitchen lines; and simply lost. We are out of work, not for lack of desire to work, but for deliberate shutdown of vision by the Eastern Establishment to understand that, as labor goes, so goes the nation and the world.

Our cities are being destroyed from the shutdown of America's productive labor force. It is becoming a dirty word to even say the word "production" because the powers that be are pushing service economies. The skill level of our potential labor force has dropped to an all-time low.

Unless we return to the principles of the Founding Fathers' understanding of the importance of labor, such as Alexander Hamilton and Henry Carey, who knew that the real wealth of any nation is the productive capabilities of the individual citizens to do the Good, we shall be returned to the status of merely a forced-exporting colony, with no ability to determine our fate for the Good of the nation and the world.

We must win this war for Economic Justice!

## Erma Craven

*Executive board member, National Right to Life Organization, Minneapolis, Minnesota*

For several decades I have waged a war to stop the genocidal policies of abortion, forced sterilization and euthanasia, being imposed against the Third World population abroad, and particularly the "Third World" population in America, which is growing at a rapid rate.

I have long documented the fact that these policies of so-called "family planning" were really plans for "family extinction" by those who were determined by the Establishment elite to be "useless eaters." This deliberate and cold-blooded murder plan can best be confirmed in the alarming policy document known as "Global 2000," which calls for the reduction of the world's brown, black, and yellow population by one-half, by the year 2000.

As this planned extermination of over half of civilization has gained momentum over the ensuing years of economic racism, amplified by the shutdown of U.S. farms, U.S. manufacturing and industry, the medical profession, and a commitment to advancing the development of moral, traditional family values and belief in the "Right to Life," I fear without a new, grander initiative to undo these evils, now destroying a belief in the primacy of life, humanity will become an extinct species.

We need to establish a New Order for Justice for the besieged populations of this world. We need to give our children hope in the value of life. For this, the Schiller Institute has my commitment and determination to fight, side by side with you.

My prayers are with you in all deliberations.

# DEVELOPMENT IS THE NAME FOR PEACE

## H. Davis Wall

*President, Local 7430 United Steelworkers of America; vice president, Charlotte, North Carolina Labor Council, AFL-CIO*

While American heavy industry, steel, machine tools and service equipment in particular, lay devastated by "post-industrial" economics, organized labor has justified its ever-weakening position by saying that it had no responsibility in determining or influencing where and how industry invested the profit incurred from labor's efforts.

This foolish policy has now come home to roost. Yet even now, the leadership of labor refuses to confront the sources of its problems of declining numbers and influence, choosing instead to believe that these can be restored by "organizing the unorganized" and moving into the service sector, thus mimicking the insane policy of management.

The leadership of labor has foolishly adopted the propagandistic notion that "cheap foreign labor" and "government subsidy" are to blame for our industrial decline and growing underemployment and unemployment. Here again they are swallowing what is being spoon fed to them, and refusing to look at the facts and institutions determining the situation.

The time has long since come for the leadership of labor to reject the axiomatic assumption that global economic trends are beyond our control or concern, and must simply be adapted to as best we can. The realization must be made and acted upon that industrial management's abrogation of its responsibility not only to maintain, but to continually improve industrial productivity and technology, has by default placed that responsibility upon labor.

I need not reiterate here the evils of the IMF or the economic implications of technological progress, that others can explain far more eloquently than I could ever hope to. Labor must nonetheless adopt these policies of practice that encourage industry and government to embark upon programs of technological innovation and industrial redevelopment. In short, those programs outlined by Mr. LaRouche in his recent message to labor on the occasion of the IBT returning to the AFL-CIO.

If labor fails to take such necessary steps, they will be just as deserving as government and management for the overflowing wrath and anger of rank-and-file membership, and will no doubt share the same fate. No matter how deserving leadership may be of the fate that nature has in store for them, it is up to those of us who have retained our sanity to see that that fate is not meted out upon the common workers, whose only fault is in having been victimized by a leadership that is more concerned with preserving its place in a dying order rather than bettering the human condition. For without them, neither industry, labor, nor our nations can hope to survive.

I send you my greetings and hopes for a productive, successful, and influential conference. Please know that although other obligations prevent me from being there with you today, my prayers and hopes are with you and our shared goals and aspirations.

# RESOLUTION

*Passed on March 27, 1988, at the Schiller Institute's conference, in Cologne, Federal Republic of Germany.*

The 240 participants at the Schiller Institute conference, Towards a New, Just World Economic Order, held from March 26-27, 1988 in Cologne, West Germany, extensively discussed the ongoing threat to Panama's sovereignty, and, after hearing the following telegram to the conference from Berta Torrijos de Arozemena, Panama's ambassador to Spain, voted unanimously to adopt its text as a resolution:

On the occasion of the Conference, Towards a New World Economic Order, I would appreciate that there be read out and considered, this personal message.

In these dramatic times for Latin America, the problem which has now arisen in Panama, has two aspects: first, that a focus for instability has been generated, which is very dangerous for the Caribbean, and second, that the U.S.A. is laying down certain guidelines for action, which create a grave international precedent. A limited view of things taken by certain strategists in the U.S.A., tries to make us believe that strategic security makes it permissible to leap over any form of legality, any legal barrier or ethical limits. The U.S.A. has promoted against Panama, a campaign of psychological action worldwide, to strip the state and some of its leaders of their prestige, and then, to launch a shameless intervention against the structure of the state itself, subverting it and destabilizing it, in a way which very possibly has never been tried before in the civilized world. The U.S.A. is in breach of the Organization of American States and United Nations charters, which imperatively describe as illegal and contrary to international law any form of economic pressure. We are faced with one of the most irresponsible acts in international politics in recent times. All nations on the face of the earth must protest against these acts by the Washington administration, because the juridical security of all nations depends on that.

# I

## The Fight for a Just New World Economic Order

# The History of the Fight for the New World Economic Order

Four decades ago, when I was a student in Europe, we would meet, filled with the radical missionary zeal of youth, wanting to see the old world go spinning down, as the poet said, spinning down the ringing grooves of change, to a new economic order. We believed in those days, and we still do, that life is the fundamental essentiality of natural processes. Life has to come into existence, life has to survive, life has to grow, life has to develop. We were sure then, and I am sure now, that the assumptions of present existence are unacceptable. One only has to regard the madness that is going on outside. . . .

I must pause and ask you to bear with me while I remember those who have fallen in this fight for a just, new economic order. There are many of them, but in particular, I want to remember Indira Gandhi, Gamal Abdel Nasser, and Marshal Tito, whom I met, as pillars of what we call the Non-Aligned Movement. They're dead, and, as the poet said, "they shall grow not old, as we who are left grow old. Age shall not weary them, nor the years condemn, as the going down of the sun, and in the morning, we shall remember them."

Fred Wills *is the former Minister of Trade and Foreign Affairs of Guyana. He is a member of the board of the Schiller Institute. Wills chaired the conference proceedings.*

And in remembering them, I shall ask you to rededicate and recommit yourselves to the purposes of this conference, because it is clear what is happening. It is clear what the solution should be. The only problem is the will to implement it. That mankind has to intervene is obvious. The quality, nature, promptness, and efficacy of that intervention is what we are met here in Andover to consider.

There are three outstanding events I want you to recall, that happened after the Second World War. Generalized technological advance took the form of the nuclear age. That was very important. Secondly, there was a proliferation of new states in the world. I give you one example. In Africa, there were three independent states in 1939. In 1987, there are now 50. There's an immense balkanization of Africa, and, by and large, they created not nation-states, but what I called state-nations. They deemed you to be a state, and therefore, a nation. The third thing, and that's why we are here, and these are all connected, is the establishment of the Bretton Woods system, in a little place in New Hampshire, where they met in 1944, and concluded treaties by 1946.

I wanted to tell you this, that being a living part of the decolonization process, one of the terms of the British, and French, and Dutch, and Belgian withdrawal from colonialism was, that you must join the Bretton Woods system. That was one of the terms. You did not get independence unless you agreed to do that. It is one of the silent, unspoken premises of alleged independence.

We were filled with hopes. We were warned. I warned, that you have to be careful and let decolonization not be the transfer of a kind of pseudo-sovereignty from a metropolis in Europe to some institution. Because—I think the French have a lovely expression, *plus ça change, plus c'est la même chose*— independence could be one of the greatest examples of nominalist hypocrisy you could ever meet.

But, there were those, then and now, who think that you are a doomsayer. Believe you me, ladies and gentlemen, the "I told you so" guy is never liked. One of the reasons why my good friend Lyndon LaRouche is not appreciated in America, is because every time a new tragedy happens, he's the "I told you so" guy. . . .

## The Goal: Development

We had ideas. We wanted a new international order. We wanted to diversify agriculture, put in infrastructure: health, housing, education; industrialize, bring the Third World— which is very important, the developing nations, 80 percent of mankind lives there, you know—bring them out of the 14th and into the 20th century. Grandiose ideas. Independence was not an end in itself, but only a means to an end. The end was development. Nations and peoples have to come into existence, have to survive, have to grow and have to develop. Those were our aims. . . .

My friends, I want to suggest to you that the test of any monetary system is its capacity to enshrine institutions of credit. That's the test. It's not whether the U.S. dollar is doing well as against the yen, or what they're doing to the German mark—that's a case, in my view, of economic murder, what they're trying to do. The important thing is, what institutions of credit do you enshrine? What access do you give to those institutions? How do you distribute credit? That is the key thing about any monetary system.

It is not whether only the dollar is monetized in gold. It has nothing to do with that. The original sin of the Bretton Woods system, I found, the original sin, was that the vast majority of mankind, in the developing nations, 80 percent of mankind, were told that they could have access to international credit only at the price of the *surrender* of their sovereignty in determination of economic policy. . . .

But we were sovereign, we thought. Sovereignty meant sovereignty. And then I had a phone call in 1971 from a secretary in the ministry of foreign trade, which hat I was asked to wear, by a heartless prime minister. The phone call said, Richard Nixon, Richard Milhous Nixon, to be quite precise, has taken the dollar off the gold standard and "let the dollar float." It used to be $35 an ounce. It shot up to $800. What that meant, of course, was massive devaluation, because we were at the time two Guyana dollars to one U.S. dollar. That was the rate of exchange.

By no act of our own, by a mere imperial order, by executive order of Nixon going on television, the money of Guyana and all the Third World was devalued. That's the meaning of sovereignty. And there were further devaluations as you went along. By the time he was finished, and the IMF was finished, as I stand here now, it was two Guyana dollars to one U.S. dollar. It is now, officially, 20 Guyana dollars to one U.S. dollar. And unofficially, in the underground economy of dope, prostitution, gambling, and what have you, it's 30 Guyana dollars to one U.S. dollar.

They have banned the importation of milk, so they say, to save real foreign exchange. That is what is happening. I don't say this because I believe in evocative phraseology. I say this to remind you, those of you who prefer to live in a cuckoo world, who prefer to think that "that's not here, that can't happen here," that those are "basket cases" down there. Let me tell you something, you must tell me, what is the difference between America and the Third World right now? Because, the American debt that these Reagan guys have caused, will have forced all of us in the Third World into the arms of the IMF. . . .

Money is a political creation. Don't get manipulated. Every time you want to present a political argument, an economic argument, you're told, "Do not criticize economic processes. That's the economic problem. Have a separate panel on that. Have some expert"—some denizen out of the dusty tomes of Ricardo or Adam Smith, or what have you. I have always believed, and I'm a hard case, and I will die believing, that the important thing about economics is its political equations. . . .

The idea of the Bretton Woods system after the Russians had decoupled and gone their way with their own machinations, was that the U.S. dollar would be the only money quoted in gold, and we would all be quoted in U.S. dollars, and institutions would be set up, IBRD [the World Bank], IMF. . . .

The important thing is, that the IMF handled credit systems, and there was bound to be a balance of payments disequilibrium, between export receipts and what you imported, and your total need to fill the gap. You needed credit, and the IMF controlled that. A kind of corrective colonialism—more calloused than the colonialism that preceded it. You all went to the IMF, and you were subjected to conditionalities. "Devalue; cut government spending"—you've heard this before—"fiscal discipline; untrammeled free market." In response to which I once told an American ambassador, that I must check with my ancestor, because he seemed to have suffered terribly from an untrammeled free market system. He was a slave.

Don't ask for all the sophisticated technology that Lyndon LaRouche has been speaking to us about. Take the California maxim. Do your own thing. Burn peat. Burn mud and wood. Import cosmetics. Don't import medicines. Things of that kind you were told.

So I reminded them again, that slavery itself was appropriate technology. And we had enough of that. It fell on deaf ears.

I was marked on by my very good friends, the British, as a person who's never satisfied. Well, I'm never satisfied, so long as a single human being has to live in degradation. So long as we go spiraling down the corridors of ruin, I shall not be satisfied. And that's why I am here.

## The Non-Aligned Fight

So what did we do? We said we'd use our forum in this fight. We'd go to the OAU, the Organization of African Unity; SELA [Latin American Economic System] in Latin America; we didn't go to the OAS, for obvious reasons; CARICAM in the Caribbean, of which I was a member; the Non-Aligned group. . . .

It was a consequence of that, that 1976 became a crucial

year for me. The year started off with Marcos holding a conference in Manila. We all went down there, and gave speeches, and I was asked to see the American government. So, being a difficult guy, I went 'round the world. I left one way through London to Manila, and I then came to California. I remember that well, because snow fell in San Francisco, and it hadn't done so in eight years. So obviously something was afoot. Then I came to New York.

And then I went and met a guy named Henry Kissinger. You've heard that name, I suppose. We seem to have in the modern era in America a vast number of educated but unintelligent men. It's a phenomenon. You can't deny that in a formal way, they have been educated. But when you examine their mental processes, their noetic processes, it's amazing. We were discussing why Guyana doesn't vote with America in the U.N. He wanted us to vote a certain way; we did not. I said sovereignty is sovereignty, and he cut off aid to Guyana and Tanzania and some other place. . . .

Kissinger greets me and starts to discuss architecture, if you please. And then he starts to discuss music. I follow him. And then he starts to speak about the Periclean Age in Greece, and I follow him, but I thought, this guy is controlling the agenda. So, I must now wrest the agenda from him. Because he's trying to impose his will on me. Understand in that kind of diplomacy, he who controls the agenda, controls the interpretation of current reality. So I said, what do you think about development in Zambesi culture? I knew he knew nothing about that. And the minute I said that, he got back to the question. He got back to why I was there. Because he wasn't going to let me set the agenda. But you have to know how to deal with lunatics of this kind.

He said to me that America loves the Third World. And that, if we only gave America a chance. I said, yeah, but who's speaking for America? Not you. I've been speaking for America more than you have. He says, I should be telling you that, to give America a chance. In my position in the Third World, you can't tell the secretary of state, he should give America a chance. I did. . . .

So I went to the U.N. and I asked for an International

8

Fred Wills

Resource Bank. . . . In my hotel room, there were quite a few supporters, and a gentlemen, who had an amazing capacity for cutting out the c- - -, and dealing with reality. And he, whom I'd never heard before, visited me and was speaking— it was Lyndon LaRouche. I listened, and I wanted to know, how could the Americans indulge the luxury of not having a man like that in their government. Because I tell you this, some of these guys couldn't get elected in the Third World.

## Call for a Debt Moratorium

I went to the U.N. and I asked for the International Development Bank, I asked for a debt moratorium. I asked for a rescheduling and restructuring of debt, a program. I was told I was a lunatic. My friends abandoned me. The Russians called me pro-American, and the Americans called me pro-Russian. The Non-Aligned group thought I was too big for my britches. My president thought I had embarrassed him. So I gave him a book on dialectical economics to read, and he said he read it. I have my doubts. He had a nominalist problem.

I knew then that it was a question of time when I'd lose my job. My problem was how to maximize my actions, while I lose the job. I gave them a good run. I enjoyed it.

We failed. We were defeated. We were defeated by the politicians, and by a monetary priesthood, I call it, disciples of Keynes, Schacht, Harry Dexter White, who is a joke, who was really Keynes's right arm. At Bretton Woods, we had two plans, the Keynes plan and the Dexter White plan. In a surge of American patriotism, they refused the Keynes plan and took the Dexter White plan, not knowing that was also Keynes. That is how international politics works. Always get the other guy to think it's his own, when it's yours. . . .

I want to say to you today, this. If this planet is destroyed, this planet Earth, it will be because of mismanagement of economic science, and not a mismanagement of physical science. It could be destroyed by mismanagement of economic science, and that is why we are here.

I want to say to you, that bad economic policies have led

9

to highly suspect accounting feats, damage limitation. That's what they're doing. But we have now transcended the possibilities of arithmetical illusion. You can juggle books, you can work your symbolisms, but with starvation, disease, and hunger rampant, those tricks no longer work. The time for palliatives, for exhortatory, verbal panaceas, is over. This is the time for fresh procedures. And that's why we're here. This is a time for surgery, and that's why we are here. Life on this planet is too valuable to be left to the idiosyncrasies and foibles, embellished and nominalist idiocies of the contemporary economists—Baker, etc. They have failed. They and their policies must go.

It is time to return to the fundamental appreciation that money and monetary systems are the servants of humanity. They are not ends in themselves. It is time to appreciate that we have reached not only a crisis in interpretation of reality, as I said before, but that we need a qualitative change in the financial systems of this world. We need new credit mechanisms. That is why we are here. We need new mechanisms for the generation of human survival. New mechanisms for the generation of human growth. New mechanisms for the generation of human development. We need rational avenues of access, by everybody—OECD, developing nations—to all the new credit mechanisms we might create. The need for intervention is clear.

## Action by Reason

We cannot remain passive in the presence of impending catastrophe. Our intervention, as exemplified by this conference, must be based on reason. The choice of action by reason, over action by feelings and appetites, is not a mere choice of options, a word they like in the halls of leadership these days. When reason is ignored, the forces of nature serve up terrible reminders. Just re-read the history of the 14th century.

We must have confidence in new technologies, that illuminate man's horizon at the moment, and all development, and the pathways to development.

# Fred Wills

Starvation, famine, disease, recession, depression, are not culturally ordained. They are not the permanent pillars of cultural relativism. Stop going down to the Third World, where 80 percent of mankind lives, and tell them, that is your way of doing things. Stop that! Stop carrying Racine and Moliere. Carry Pasteur and Madame Curie.

We must not and cannot allow the growth to degenerate into a pandemic wilderness. I say we cannot furl the flags of human progress. That is what this conference is about. Once again, we are met to establish a new monetary system. We are not here to repeat the mistakes of the old. Above all, we're not here to deify, hallow, the assumptions of the old system. We are here because we know that mankind need not remain passive in the face of impending doom.

But if we don't do it, there's nobody left to do it. We are the thin, red line, standing on the lip of progress. If we lose, it's not personal battles and personal tragedies. It's a decision about whether mankind on this planet has a future. I therefore ask you, in the name of those present, in the name of those departed, who've been in this fight for a new economic order and new monetary systems and new mechanisms of credit, I ask you not to let us fail again. True, we've come back to the Northeast. Bretton Woods is in New Hampshire, and we're in Andover. But that's a mere nominalist coincidence.

I think we have the intelligence in this room, the leadership and the directionality, and the resolve to triumph over the present.

LYNDON H. LAROUCHE, JR.

# The Tasks of Establishing an Equitable New International Monetary Order

I wish to put into focus the feasibility, not merely of adopting a design which would address the need, but the feasibility of implementing that design, effectively. Now, in part what I shall say, I have said, and will be broadcast this coming Thursday on a CBS nationwide broadcast.

I repeat that here, not because it's in the broadcast, but rather, I put it in the broadcast because, at those prices, it's the most important thing to say.

The United States is on the brink of collapse, not merely economic collapse, financial collapse, but *national* collapse. We are very close in many parameters of sovereignty, to becoming a Third World nation, as exemplified by the fact that, that half-Baker, in the Treasury Department has proposed to issue U.S. bonds—U.S. debt—denominated *not* in dollars, but in deutschemarks, yen, and so forth.

Once the United States were to denominate large portions of its debt, national debt, of government in foreign currencies, under conditions of a collapse of the exchange value of the

Lyndon H. LaRouche, Jr., *a Democratic presidential candidate in 1980, 1984, and 1988, is a founding member of the Schiller Institute.*

U.S. dollar, the United States becomes a Third World nation, in all but the final result.

And around this country among over 50 percent of the U.S. population, *not only has the poverty* been increasing for approximately twenty years, but I can show you in the United States actual Third World conditions on a large scale. I can show you cities, and portions of our cities, which look like bombed-out cities in Western Europe at the end of World War II. And, I can show you Americans who often become insane by the conditions under which they live. Americans who struggle for subsistence, and compete with the rats and cockroaches which vastly outnumber them in these places.

Oh, we have an image of our so-called "ghettoes," Hispanic and black ghettoes in the United States, but that's not the extent of poverty. I'll take you across the line to New Hampshire, and everybody in New Hampshire in government, and the federal government, will tell you New Hampshire is a fine place to live—very prosperous. This is based on the report that there's a very low percentage of unemployment in the labor force.

Well, you have to look at two other things—three other things. First of all, you have to look at the pay of the people who are employed, relative to what it costs to live. It takes two non-breeding, mated pairs of yuppies to sustain the acquisition of one house—recently constructed, which shall not outlive the mortgage. That is not my view of prosperity.

I can show you the majority of the population of New Hampshire is objectively in worse economic condition today, than I saw first-hand during the 1930s! I can show you that the problem is concentrated largely among senior citizens, of which New Hampshire has a high percentage in its total population. The reason is that the young people of New Hampshire got out of the state, because there were no opportunities there. And what have moved in, are the yuppies who came in to follow the search for cheap labor by industries out of the [industries located on the Route] 128 complex, largely.

I can show you a state, New Hampshire, which lacks basic economic infrastructure. The entirety of New England is now generating a peak of about 18 gigawatts of energy; the con-

13

sumption of energy under depressed conditions and cold weather up here is 18 gigawatts. The region is *losing* energy capacity through attrition, but you could not put up industries here, in New Hampshire, or in northern Massachusetts, to expand opportunity. *The infrastructure does not exist.* The energy doesn't exist; the transportation doesn't exist; services don't exist; the medical services, the school services, and so forth don't exist!

Like the Roman Empire—Italy—in the last phase of decay before it collapsed, before the barbarians moved in. I hear they're gathering in Vermont. So we are in that situation.

The United States has embarked on a strategic policy with a President who is under the control of a friend of Armand Hammer, i.e., his wife; which means that Western Europe, under present policies and trends will become an extension of Finland at a rapid rate. Any other interpretation of the INF treaty, and associated agreements is absolutely absurd, even though you hear it from many sources. Anyone from Europe, from the inside, who knows the situation, understands the logic of what's called the "Finlandization" process in Western Europe.

The great patriot, Reagan, has set this fully into motion, and has attempted, and is dedicated to making that trend irreversible before he leaves office, by succeeding the INF agreement with the START treaty, which, essentially, would ensure that the Soviet Empire would dominate the world—irreversibly—for a long time to come, beginning in the 1990s.

Under those conditions the United States would become a client-state of the Soviet Union, unless it resisted that status, in which case the United States would be destroyed—unless it could win a war—in isolation from its former allies in Western Europe, Japan, and so forth. And the world would go under Russian conditions—and don't have any illusions about Russia.

## The Russian Empire

Russia is a modern caricature of the empires of Babylon—of the Achaemenid Empire, of the Roman Empire, of the

14

Byzantine Empire. Russia is the empire, first of all of a master race, the great Russian race. What is called the Soviet Union, is a collection of "captive peoples," mostly of Turkic-speaking origins, who are subjected to Third World conditions. The rates of mortality, of infant mortality, and other conditions inside the Soviet Union, in the Turkic populations, compare with those of any average Third World country which we consider oppressed.

This oppression is imposed by the Great Russian master race! Of the Third Reich of Russia, the Third Rome. Outside of Russia, itself, we have the satrapies, the colonies of Eastern Europe. In the colonies of Bulgaria—a friend of ours was recently there—in Bulgaria, there is real misery; in Romania, it's worse; in Poland, it's worse; and in East Germany which has about the same cultural level as Western Germany, when they want to celebrate, they cover the fronts of houses with grey paint, which peals off very quickly.

Why is this true in Eastern Europe? "Oh," someone said, "there's communism." That's double-talk! The reason is, because *that's the way the Russian Empire rules the world!* It comes in, it tells its subjects, "You cannot do this, because you will compete with us. So, you must be on a lower level than we are. You must subsidize us, by supplying us with what we wish to buy, which we will purchase with the credit *you will give us!*"

That is what is happening in Germany, now, in the increase of East bloc trade. All Western Europe will be subjected, if this occurs, to Eastern European rules of the game, increasingly. And, there are those in the United States, including friends of Armand Hammer, Dwayne Andreas, and others, who are prepared to put the United States through the same process.

So, we stand at a point where the United States is at the verge of not only ceasing to be a world power, but of becoming, if it peaceably submits, at best, a client-state on the outer fringes of the Soviet world empire. And, under those conditions *there will be no development.* Under Soviet world rule, the conditions of life of the so-called Third World will be far worse than they are today.

15

## Development is the Name for Peace

What's at stake is not only the United States, as a sovereign nation, not necessarily a sovereign world empire, but a sovereign state. And what's at stake, if this process continues, is the very existence of humanity. Because, as we knew many years ago, because of the simple laws of epidemiology, that if the trends, and conditionalities, which were set into motion between 1967 and 1972, were projected further—as trends—that you could calculate the effect of these conditionalities upon the per capita level of existence in certain parts of the world. And so, my friends and I did those calculations back in 1974, and on that basis we projected that by the middle of the 1980s several things would happen. And we focused, in particular, on developments in the Sahel in Africa, where we foresaw the worst effects to break out first. We forecast the cholera, typhoid, etc., epidemics to reach major proportions, and also predicted that a major new disease, including some kind of pandemic, previously unknown to mankind, would erupt as a mass planet-wide killer during that period, as a result of the breakdown in economic conditions.

The world—Africa, for example—is being deforested. India has been deforested, with catastrophic effects on its climate. The deforestation of Africa, the same. Why the deforestation? Because we don't allow them to have energy supplies or alternative kinds of fuel. The poor people cut down the trees for fuel, to cook their meals. Why? Because we say, "appropriate technologies"; because we say, "You cannot have modern energy technologies."

If this continues—particularly with the HIV virus, and its eight *now known* mutations—under these conditions of economic decline, and spread of pandemic and epidemic diseases all intermingling and interacting as co-factors of one another, we are at the point where it is possible to project the certain extinction of the human species by some time during the first half of the next century, perhaps even the first quarter. That is a very real prospect before us.

And governments are lying; the World Health Organization is lying about this. *They're not mistaken*, there's no honest difference of opinion. *They're lying*. AIDS alone—what's called AIDS—alone, can be transmitted by any possible means that

16

any virus can be transmitted: You simply require the right conditions and you may have to wait a few weeks before the virus evolves, or adapts itself—adapts its outer coat—to find a new opportunity. If we were to fight the disease, as we could, this would mean spending, in the United States, for example, in the next year, $50 billion. It would mean, very rapidly, an expenditure of $100 billion. It would mean within four to five years an expenditure of $200 billion annually just to fight this disease. And, the Reagan administration says, that to expend that kind of money in the face of the current budget crisis, would be contrary to the administration's economic ideology, and therefore, we are going to lie, because we are not going to let the people be aroused into forcing us to spend that kind of money.

So, we're at an existential point where the question of a new monetary system—a new economic order—is no longer a question of choice, it's no longer a question of abstract morality, it's no longer an ethical question, as we define the word "ethics" in vulgar use today. It is a question of whether the human race does, or does not have the capability of making those decisions, which constitute our species' moral fitness to continue to survive. It is not an abstract question of justice. It's a question of human survival of us all, and of the grandchildren of the coming generations.

The decision will have to be made soon. For various reasons, the decision will have to be made inside the government of the United States. There is no alternative.

Granted, the industrial economies of Western Europe, in total, represent today a significantly larger economic potential than does the United States. Japan is a much more powerful economy, than any other economy in the world, per capita, today. And one could say that if the United States fails, some combination of Japan and Western Europe might appear, which could take the place of the United States in starting a new economic order in the world. Politically, that's impossible.

There are people in these various countries, in Japan, in Western Europe, people who are very positive, people who will respond. But, none of these countries has the capability of pulling together those forces, in a united way, sufficient to save humanity, and the Russians won't allow it. Only in the United States, and the United States government, do we have the means, not to solve the problem, as such, but the means to make certain decisions, which will bring about the kind of coalition of forces needed to make the change effectively.

I indicate the present situation. The present monetary system essentially came to an end by about 1982. I was there, I was consulting with the Reagan administration, in pushing what became known, a year later, or so, as the Strategic Defense Initiative. In that connection, I warned the Reagan administration, through the National Security Council, and other institutions, that as a result of decisions made at the end of 1981—international monetary decisions—that the external debt of the nations of South and Central America was about to blow out, with Mexico at the head of the list. I warned of that over the first six months, and after meeting with a gentleman (who should be here, but he said, "The world would blow up" if he came here—the former President of Mexico, López Portillo) in a discussion of the situation. I had reviewed to him what the problems were: that we could expect the Mexican debt situation to blow out by September of that year, 1982, and that the forces in the United States, were prepared to take Mexico apart piece-by-piece, a process which has been going on ever since, and which is not completed.

## Operation Juárez

The next big destruction of Mexico is about to occur very soon, by the self-destruction of the leading party, the PRI. Balkanizing the political processes, possibly turning the north of Mexico into a province of the Anglo-American drug pushers, who have taken over pretty much as they're trying to take over Colombia, and then divide the country, and turn it into the conditions of civil war. So, I indicated to him that we would have to act very soon, not in the case of Mexico, but in other parts of the continent, to reverse this process if we were going to save these countries, because *all* of them were doomed similarly, on the basis of the policies floating around the Reagan administration at that time.

So, in that context, friends of ours, including friends that Fred [Wills, former foreign minister of Guyana, who chaired the conference] just referred to, the SELA [Latin American Economic System] group, approached me, and said I should put my ideas into a book-length manual—stating my ideas to all the people, particularly in Hispanic America, and all of those who agreed with us—to give them a working manual so that they could work together to common effect. There was a very significant movement in that direction, at that time.

When I completed the project on the first of August, I presented the manual to the Reagan administration. About two weeks after I submitted the manual, of course, the Mexico debt crisis fell, and the entire world monetary system nearly went over the cliff in a two-hour period on the day of the Mexican announcement.

The President of the United States, Reagan, called President López Portillo on the phone and offered to use U.S. credit for the United States to help Mexico carry over this particular crisis—that delayed the crisis. López Portillo, with, at that point, the commitment of the President of Brazil and the government of Argentina, acted to implement a set of proposals identical to those which I had outlined in my report, which I entitled, *Operation Juárez*.

There was a fight inside the Reagan administration, with people inside the National Security Council, CIA, and elsewhere, taking my side on the issue, and Henry Kissinger's friends, and Kissinger Associates—Donald Regan, and from outside, Walter Wriston, in the New York banking community—taking the opposite side.

Needless to say, we lost the fight. The President of Brazil chickened out, betrayed the President of Mexico. The Argentine junta demonstrated what kind of a military leadership it represented by chickening out, and betraying the President of Mexico. President López Portillo was left hanging out to dry, and his country was chopped to pieces, piece-by-piece. It is now at the point of virtual destruction.

Compare Mexico in 1982, with Mexico today. You say,

## DEVELOPMENT IS THE NAME FOR PEACE

"Here's a country which has been destroyed!" Just as much as if a Nazi occupation force had occupied it during the middle of World War II; that force would have done no worse than has been done by a government which has carried out point-by-point, *nothing but the orders given to it* from London, New York, and similar locations. . . .

What happened is, as a result of that, President Reagan took action, together with the New York banking community, which resulted in creating the biggest John Law-style financial bubble in history. That bubble kept going on. The U.S. economy collapsed. There never was an economic recovery in the United States. Don't believe it! The President is stupid on these questions, so I can't accuse him of lying. On economics, he's insane, clinically insane, always has been, ever since he got into political life. But, he's been saying, "59 months of economic recovery." We had the biggest financial collapse, since Black Friday of 1929. White House reports came out the next month: Sixty months of unbroken, uninterrupted economic recovery. This period of 62 months of so-called "economic recovery," since 1982, is what he dates the economic recovery from.

Through the entirety of this period, what has happened is, U.S. agriculture has collapsed, U.S. industry has collapsed, U.S. industrial employment has collapsed. The average level of real content of the per capita family market basket has collapsed; infrastructure has eroded, and collapsed; the purchasing power of the dollar on the world market has collapsed. The President calls this "recovery." He must be standing on his head to read the charts.

What grew? Yes, something grew. And, Reagan had the figures every month: Admittedly, the figures were fake. Since 1983, virtually no figure by the U.S. government has any correspondence to reality. We had a trade figure recently: completely fraudulent. We had a GNP figure: completely fraudulent this month. The government has simply made up the statistics reported as the official reports for political purposes, with no regard to what actually happened.

But one thing did grow. What grew is what's called, "value added from financial revenue sources," the value added of

20

finance. When the real economy is collapsing, and the nominal value of financial assets is increasing, what are you doing? This is called, generating a "John Law-style financial bubble." And, last October that bubble began to collapse. It is a bubble— the magnitude is between $15 and $20 trillion internationally. It is a financial system no one could bail out, even though Reagan and Bush are trying. It is going to collapse. *The collapse is inevitable. It is unstoppable.*

## Reagan's Delusions

The reaction to this collapse is that President Reagan says, "There is not going to be a collapse while I'm in office. I've got to go out as a man of peace and we'll let the Russians take over afterward, let the depression occur afterward. But, let me go out as a man of peace. Let me go out, and go to my death, or whatever it is that I've got—let me go out with a grand illusion. Let the film close with Bonzo a hero."

And George Bush says, "Yeah, man! I've got to be the next President, and I think I'd have some difficulty running as Herbert Hoover. So, do anything. Sell children into slavery; beat up 15-year-old children—whatever you have to do—to delay the crisis until after November of 1988. Then, let it all hit, because I'll be President!"

Great fellow, that Bush. Contrary to the image he presents as a simpering preppy, underneath that image there is a real down-to-earth George Bush—a real knuckle-dragger—as you saw on national television with Dan Rather. This guy's a thug, essentially. That's the situation. The situation is worse, however, than merely the idiocies of a senile President, and a George Bush—you will never notice when he becomes senile, because there will be no change. His talents lie from the neck down.

What has happened is that, since the outbreak of the events of early October to middle October, the President, the leadership of the Congress, the Federal Reserve Bank, the Federal Reserve System, the leading U.S. bankers, the leaders of the political parties, and most of the institutions, have been doing and saying exactly what Herbert Hoover, the head of the Federal Reserve System, the head of the Treasury, the leader of the Democratic Party, other leaders of the Congress, the New York and Boston banking community, did and said between 1929 and 1931.

In Europe—except for some noises out of France, [Finance Minister] Balladur and [Agriculture Minister] Guillaume—what we're hearing from Europe is exactly the same policies, identical, virtually word for word and identical in substance. The same thing that was said between 1929 and 1932. The result of this is as follows: The crisis we're in, is immediately a financial crisis associated with a collapse of a gigantic financial bubble—a John Law-style bubble. In the 1920s, the bubble was the hypothecation of a structure of French and German debts to the United States, on the presumption of the Germans' payments of the war reparations debt. When the point was reached of the Young Plan, that it was obvious that the German war reparations debt could never be paid on those terms, the markets responded to this happy news by collapsing. However, the bubble, the Versailles bubble, which set off the 1929 to 1932 collapse, was relatively, as well as absolutely, much smaller than the financial bubble which has been built up over the past twenty years since Lyndon Johnson began to take the system apart.

Therefore, what we face is, in many respects, a repetition of the 1929 to 1931 developments, with two general exceptions and one special one. First, the process is much deeper than during 1929-31; secondly, the tempo of the process will be more rapid than 1929 to 1931, which means that, at the present rate, we could expect to be in the depths of a depression much worse than 1931-32, by sometime in 1989 at the latest. This will be the greatest catastrophe in the modern history of the United States, if it continues. Now, the third problem is: The political parties of the United States, and the quality of government, are vastly inferior, to the quality of the political parties and government back in 1929-32. And, the quality of the population generally, in terms of educational level, in terms of the stability of institutions of family life, in terms of resources to fall back on under conditions of mass unemployment, are far poorer than they were in 1929 to 1932.

Therefore, we're going to have to make decisions very quickly, because the combination of what is happening on a global scale and strategically with it, the rapidity of this crisis inside the United States, means that we are at a point of irreversibility—a *punctum saliens*, of which we either make the necessary decisions, or we can sit back on a mountain top, if we can get there, and contemplate the great spectacle, the greatest of all Roman circuses—the death of the human species, or at least of civilization, as we know it.

And, therefore, unless we can find a President of the United States, who can, as a candidate, begin to shape the events of the coming months and who can assume office in January of 1989, I think the chances of humanity as a whole are grim ones for a long time to come.

Now, I'll indicate the more positive side. The nature of the crisis lies not with the objective problems we face. The crisis lies essentially with the fact that we haven't got, in our governments, the brains to respond to objective problems with available objective solutions.

## What President LaRouche Can Do

Just to indicate what I would do as President on the day of inauguration, and I don't think that there will be much that will change in the meantime to cause me to adopt any different measures or require any measures in addition to those I would envisage now. They're not too difficult, you just draw up the list, and when you're inaugurated and sworn in, you've got the authority to begin signing the presidential directives and sending the bills over to Congress.

Under the U.S. Constitution, the President of the United States, with a certain role contributed by the Congress, has adequate powers to deal with a crisis, exactly like the present one, with no impairment of those liberties, or the constitutional guarantees, provided by the Constitution. In addition

to the constitutional powers, particularly those under Article
I of the U.S. Constitution, the Congress over a period of time
has given the President emergency legislation, chiefly grouped
around the Federal Emergency Management Agency acts.
The agency itself, and the acts associated with it—many of
these proposals by the Congress are bad. They're bad legislation, but, nonetheless, they're on the books, and a President
who has the brains to do so, can pick from this legislation.
Simply by declaring a National Economic Emergency, he can
pick a menu of actions which coincide with exactly what has
to be done. The President can, in effect, seize the Federal
Reserve System, discontinue those practices of the Federal
Reserve to which he objects, convert the Federal Reserve
System, into a system of national banks modeled upon the
First Bank of the United States, under [President] Washing-
ton, or the Second Bank under Monroe and John Quincy
Adams.

In addition to those measures, and the use of regulatory
powers of government—exchange controls, capital flight controls, export-import controls, regulations of banks which are
in trouble to make sure they don't close their doors, regulatory
actions to defend the value of the U.S. dollar on world markets,
regulatory actions to protect the value of U.S. government
debt in the form of bonds, and U.S. Treasury bills and devaluation—the main thing the President has to do, is to know
how to use the provision of our Constitution, which has been
not much observed in recent decades.

Under our Constitution, the creation of U.S. currency occurs by a bill presented to the Congress for its deliberation
and action by the President. This bill, when passed, when
enacted, authorizes the U.S. Secretary of the Treasury to issue
a certain quantity of U.S. Treasury currency-notes, as currency. Now, what will be required over the coming two years,
in the United States, to deal primarily with the domestic
requirements of the United States is about $2 trillion a year
in issue of U.S. Treasury currency-notes. These notes would
be lent through the Federal Reserve Systems banks, which
will be functioning as national banks.

These banks, in turn, will usually lend these notes to federal,

state, and local agencies for capital improvements in infrastructure; to public utilities for capital improvements in infrastructure; for farm production loans, and capital improvements
in agriculture; for industrial production loans, and capital improvements in industry, or expansion in industry; and for long-
term to medium-term export financing of products by U.S.
exporters to foreign countries. An intelligent application of
these funds would limit their application to these categories.
That is, if you wished to go into the insurance business, you
couldn't borrow this kind of money. If you wished to set up
a casino, by no means could you borrow this kind of money.

The important thing is to make sure that the flow of these
funds does not go into administrative, sales, financial services—overhead of the economy, except in the professional,
scientific arrays of services—but goes entirely into expanding
the labor force of operatives, and their productivity.

To give you an indication of the effect of this: An increase,
even without any significant increase in technology, an increase of the number of industrial operatives—that is both
infrastructure and industry—employed, say during a three-
to four-year period, in the United States, would increase the
per capita physical output of the United States by between
20-25 percent. In point of fact, with the technologies we have,
and we'd be obliged to use, it would be closer to 30-35 percent.

## Creating a New Monetary System

There is very little that you couldn't fix in the United States,
if you started from an increase of total output of about 35
percent per capita. There is no budget that couldn't be balanced, and so forth. Take this, and put this in the context of
international economic and monetary reform. The Bretton
Woods System and its zombie relic, its Dracula relic, called
the "floating exchange rate system": They killed the old Bretton Woods System, then they brought it back as a walking
corpse, which walks at night and sucks the blood of nations—
the floating exchange rate system. That thing just has to be
scrapped! It's a very simple thing to scrap it. It's a creation

of treaty agreements of governments. If governments abrogate those treaty agreements, or alter them, it simply ceases to exist. The International Monetary Fund can sit there, it can vibrate, it can oscillate, but it just sits there. The same with the World Bank.

The monetary system has to be based on the authority of sovereign governments. It is effectively a treaty organization among sovereign governments, and has no legitimate authority, except as a treaty organization of sovereign governments as partners. Therefore, what we simply do, is we take the old monetary system, put it to one side, put it in the closet, and open the closet to horrify children on Halloween. We sort it out later.

The question is: How do we generate growth? The first thing that has to happen is, as President, I would have to have most of the so-called Third World leaders, as in the capacity of preferably either Presidents or prime ministers, or foreign ministers, or some combination, meet, and settle immediately, the question of restructuring and reorganization of debt of these nations, insofar as it involves the United States. If the United States government signs a memorandum of agreement to such effect, simply a signature on a memorandum of agreement is effectively a treaty, which the President can issue as a presidential directive in an emergency, and then pass it as a bill down to the Congress, to be treated as a treaty and make it treaty law.

But the President can do a great number of things under emergency conditions, in this form. Once the United States government, once the President of the United States, has entered into such an agreement with a group of developing nations on restructuring and reorganizing of their external debt, and expansion of their import capacity, and conditions of new volumes of loans for economic development, the rest of the world just has to go along with it. And there, we can be assured that the forces in Japan which agree with this kind of policy, would join with it immediately, and they would become predominant in Japan, as opposed to others who tend to be pro-monetarist. In Western Europe, the forces typified by the statements of Balladur, of Guillaume, would become

predominant. The crazies in Israel would simply have to go
and find themselves a new promised land on the Moon, and
the sane ones would accept what we call the "new Marshall
Plan" for collaboration with their Arab neighbors on this basis.

Of course, the developing countries wouldn't be much trouble. We might have trouble with Khomeini, but I don't think
he's going to be around too much longer.

On the basis of that, the United States, of course, would
enter into matching agreements with our friends in the OECD
nations. And, thus we would have, in effect, the basis for a
new monetary system, simply by these kinds of agreements.
What would make it a monetary system, would be the agreement of the other countries, the Western European countries,
and others, to agree to create credit, not-for-money loans. I
don't think that lending money does any good, it just leads
to usury. What should be lent are strictly lines of credit: short-
term, medium-term, long-term credit. There's no sense in
the United States government or any banks running around
giving countries money. It doesn't do any good, and usually
does a great deal of harm. The money somehow disappears
in Swiss banks on the way into the development project, in
most cases, *not* into the country.

Give these countries lines of credit, for their infrastructure,
agricultural, industrial development projects, including such
things as health programs and educational systems under infrastructure. Supply them what they need. Give them the means
that they require to employ vast armies of unemployed labor,
or *mis*employed labor.

But, in general, in my opinion, from looking at many development projects in developing sectors, most developing
countries could undertake most large-scale development projects, using 80 percent domestic resources; what they require
from foreign countries is essentially certain crucial included
elements of the project which amount to about anywhere from
5-20 percent of the total package. The trick is to enable countries to survive on their own resources, to give *them* the ability
to mobilize *their* labor, to give them the ability to lay the basis
for their own development, as sovereign states. And we can
do that.

## Development is the Name for Peace

It's no mystery for those of us who are economists, particularly the physical economy—and I suppose I could do a pretty good job right here, if we want to take the time to do it—to run off a list of major infrastructural development projects which would transform this planet. These infrastructural projects would create the domestic markets in the countries they affected for the growth and development of agriculture and industry. It would mean new industries; it would mean that increase in food supplies would come automatically. For example, railroad projects: We have, now, better railroads, we have the magnetic levitation trains if we have the power to run them, which are cheaper, better—cheaper to maintain, cheaper to build—which can run at speeds of 300 to 400 miles an hour, if you have to run them at that speed. We need now railroads, water-management projects both for transportation and for better utilization of water for general purposes in agriculture, and control of the environment. And, above all, production of power.

We know there's no escape from power production, and despite some people's sensibility, there's no escape from nuclear power production. There is no alternative. Look at the deforestation of Africa, and India, and you see the fact.

## What Is Development?

Look at India—how does it power its economy? It takes coal, runs it from the mines of the north down to the cities of India, and the movement of tons of coal by freight car is destroying the Indian railway system. Without nuclear energy India is doomed! It's not a matter of choice. There is no alternative. Yes, there's great hydroelectric potential, but hydroelectric projects, properly managed, give you very little net energy, because if you manage them properly, you use as much energy to maintain the system properly as you get from it. Or, if you get power from it, you cannot control, at will, the time you get the power from it, you have only certain parts of the year, and certain conditions, under which you get a significant net power production.

28

# Lyndon H. LaRouche, Jr.

We don't have fusion power yet; we should, but we don't. So, therefore, in this area, you can measure it with all the figures through the economic history of mankind, the level of productivity and income of a population, is a function of the density of usable energy, supplied per person, per square kilometer. The difference between India and a developing country, or other developed countries—Japan, North America, Europe today—is infrastructure measured in power. There is no development without infrastructure. It's impossible, it's a physical impossibility!

Someone says, "We're going to develop our industries, and our agriculture, rather than our infrastructure"; they don't understand economics: It's impossible! You can measure this in calories, measure this in kilowatts. The number of kilowatts of infrastructure, consumption of energy, per person, and per square kilometer, determines absolutely the upper limits of economic development in terms of per capita productivity and consumption. If you don't have that development, you are doomed to a level of development which coincides with the amount of energy per capita, per square kilometer you have.

So, in those terms, water projects, some reforestation projects, transportation projects—including rail—but particularly in water management, power, and other infrastructure, such as health systems, school systems, the development of new kinds of cities, which are cheaper to maintain, more durable—these kinds of projects—are what the world needs. It really doesn't need to think of much else.

Yes, the rest of it is easy. Once you have the infrastructure, then it's very easy to determine what industries you want to put on infrastructure. Industries are like electrical devices that you plug in the wall: They work if you have the plug, the electricity supply, into which to plug them—in this case, the infrastructure supply.

Now, this is beneficial to both the developing and developed sector. Again, our economic policy in Europe, the United States, and Japan—but particularly Europe and the United States—over the past twenty years, has been clinically insane.

The healthy development of an economy starts by decreasing the percentage of the total labor force required in rural production, to increase urban production—unless you get too many salesmen, bankers, clerks, shoeshine people, and so forth—that's insane. But development will occur under these conditions, as long as you keep the amount of administration, financial, low-grade service, and so forth to a minimum; keep your number of parasites to a minimum—you can have one parasite in the zoo to amuse the children—but generally, keep your parasites to a minimum, particularly, the ones who get very rich at that sort of thing.

Then, the urban industries grow, as Hamilton laid it out. The urban industries grow on the basis of a healthy interrelationship between the urban community, as a manufacturing community, primarily, and rural production. Urban development depends upon growth: movement away from consumer-goods production into capital-goods production. And, in terms of these ratios, the level of energy development, per capita, and per square kilometer, you can measure the absolute viability of economies, without knowing a thing about prices, without knowing a thing about money prices.

In the United States, we've been insane: We were insane throughout the entire postwar period. The so-called, "Eisenhower recovery" was a piece of insanity which lasted three years and came to a screeching halt in 1957-58. Why? Eisenhower had the theory from Burns that you had a "trickle-up" economy: If you used consumer credit to expand automobile sales, everything would be good. Insane! Insane economics, which ruined us during the late 1950s.

The trick in economy is to put the credit into the expansion of the capital-goods sector which throws off and generates technology. The demand created by the capital-goods sector *creates* the basis of the growth for the consumer-goods sector. Then, that's how you maintain full employment in an economy, by expanding capital-goods investment—and employment—to absorb as much as possible, a full labor force. In the United States, we've done the opposite. Our machine tool industry is almost nonexistent; we've destroyed our producers-goods industry, generally. Our steel industry doesn't exist: We say, "We can get steel cheaper, by stealing it from Peru, or from Mexico. We can get food cheaper than from our farmers, by stealing it from countries that are hungry," or where there is vast hunger, such as Brazil. This is President Reagan's economics.

What is beneficial to the developing and so-called industrialized countries is to eliminate, as much as possible, all export of consumer goods, except absolutely indispensable goods, such as food, when needed, in the developing nations, and almost to make a law against it, or to use regulation—export-import regulation—to prevent this from occurring. We don't wish any cosmetics going from the United States to Africa, it'll just make the Africans look ugly, and I see no point in that. Our people in the United States are ugly enough already; you see men running around with these cosmetics: It's terrible.

What we wish to export, and should wish to export, are essentially two things: It's sometimes called "technology transfer," capital goods, and certain specialized qualities of engineering services; that's all the United States should *ever* desire, to commit itself to exporting to developing nations, because if we increase the rate of development in developing nations, we have two effects. First of all, we increase the turnover of our capital-goods industry simply by more sales. And by increasing the turnover in the capital-goods industry, you actually cause economic growth in the United States—simply by exporting, even before you get money back on the goods exported. Secondly, by increasing the per capita productivity in the developing countries, well, we're doing fine, we're letting our customers grow. Now the United States is insane: Our policy makers believe today that the best way to build a market is by killing the customers, which is what they've done with the developing sector with monetary policy. The intelligent policy is to do the opposite.

What we have to reach agreement on, to create a monetary system, is to get the United States, Japan, and Western Europe, or most of these nations, to agree on a new basis for pegging currencies to fixed prices; going back to a gold-reserve standard for that purpose; to issue credit at agreed terms of credit; to have a schedule of priorities on issuance of credit; and to have regular meetings among various countries, de-

31

veloping and industrialized, to set priorities and goals for imports, exports, and investments. What governments will do, as a result of those agreements, such as the government of the United States—its export-import bank and other institutions—is to simply allot every year, for export-credit purposes, a certain percentage of a total amount of lending power to each of the categories listed, by country or by region of the world.

The way we shall operate is, the United States will become a major exporting nation again. Anything else is insane. Instead of the United States, Japan, and Western Europe trying to take in each others' laundry by selling to each other across the fence, Japan and Western Europe will be told: "No more, except in very specialized categories such as spaghetti, pasta, good European wines, and so forth—we've got to have that for the U.S. population. But in the high-ticket items, such as consumer goods—get out of it—the United States is not going to be your market anymore, for these kinds of consumer goods. You're going to direct your investment and production into providing capital goods for the developing sector. And you, Japan, we, the United States, and other countries, will come to agreed terms on sharing that market potential, with the consent of developing nations. And what we're going to export is capital goods, in order to rebuild this planet."

## The Punctum Saliens

A perfectly feasible proposition! It all hangs, of course, on making sure the next President of the United States does that. But, we have two choices. Either we don't do that, in which case, you can write off the human race. Not necessarily extinct—that could be possible—but you can write off civilization as we've known it, for a long time to come. We are now at the *punctum saliens*! The next twelve months, or so, that's the *punctum saliens*. If it isn't done then, it will never happen, at least not within foreseeable generations. So, that's the only thing we can allow to happen.

Now, as to what will happen, I don't think we, at this

32

# Lyndon H. LaRouche, Jr.

conference, or others around the world who share our concerns, should worry in the least whether what we desire to happen, will happen or will not. That is not in our power to determine. We'll do the best we can to make sure it happens, but we don't have the power to determine that.

We cannot ensure that the voters in the United States will be sane. As a matter of fact, from their recent pattern of choices in the postwar period, we find that they tend to be the contrary. They've elected a parade of prize idiots of the twentieth century, either men who are mediocrities by training, or agreed to be such for the privilege of enjoying the pomp and circumstance of holding the office: As long as they did nothing in office, they were allowed to be President—Gerry Ford's an example of that. A man who had no idea of what it meant to be President, but he sure liked the pomp and circumstance. And, as long as they didn't bother him with too many decisions, he could just go around being absolutely happy.

But, we don't control that. We cannot—facing a problem of this nature, the fate of humanity—we cannot say, "Well, we will do something about the fate of humanity, if you will assure us that the American voters are going to behave intelligently this year." Well, that seems immoral to me. My view is, that we must do what is necessary. We cannot associate ourselves morally with any enterprise, except *that which is necessary for humanity*. Therefore, win or lose, let us dedicate all of our exertions to the maximum degree to the only thing worth doing, not dependent upon whether we can guarantee success or not. I would rather die, having failed at doing the only thing worth doing, than die succeeding in contributing, supporting, or tolerating the catastrophe which is otherwise going to befall mankind.

The prospect of my becoming President is a highly speculative one, but I think I just might do it, because of the nature of the times. In crisis, all kinds of strange things, for better or for worse, happen. The prospect of finding some other candidate who might be elected, who would do it, is virtually zilch—zero. None of the visible candidates would do anything but the opposite of what I've outlined, apart from Gary Hart's saying nice things about the Third World, and being nicer to them. That's like Lady Do-Rightly handing out doilies to the poor at her back door twice a week for an hour at a time. These are the kinds of things which make charity a disgusting word. Most of them are *evil*. Dole's program is evil. Bush *will be evil*. Most of the Democrats will be evil. Nunn would be evil. Cuomo would be *much more evil*. He'd not only steal from you, he'd send a racketeer down to take it from you, with his mafia friends; and Bradley is a Rhodes scholar, you can ask Fred about what that means.

So humanity, the future of humanity, seems to be a very unlikely prospect, but as I say, we must put ourselves and our efforts to the only thing worth doing. Nothing else is worth doing. Do it right! Face each of the problems involved, both the technical-economic problems, and also the political problems, of affecting the terms of collaboration among nations, which both meet the requirements of respect for their sovereignty and also respect for the fact that their sensibilities may be different than those of some of the rest of us.

We must bring these nations together, we must bring them together on an equitable basis, we must bring them together on the basis of respect for their sovereignties. And we must bring them together with the idea, that what we agree to do is not something that's going to be served on paper, passed off to special study commissions. Those are wonderful things, those study commissions. When a government wants to appear to do the right thing, without ever having to do it, it creates a study commission, a feasibility study. When I hear "feasibility study": "Oh, we've decided to support that!" "Oh, yeah?" "Yeah, we're putting out a feasibility study." "Ah, you mean you're not going to do it, but you don't want people to be able to accuse you of not doing it." Everything that has to be done of importance, we could do right now, without any feasibility studies. So, maybe the first plank is, "It's against international law to organize a feasibility study." It might be a great boon to development! It would force a great number of politicians in governments, to put up, or shut up.

So we must come to deal with those kinds of problems. We must also, in doing that, understand the importance, particularly, to developing nations of a sense of full participation, of sovereign and equal status in the process of deliberations which we propose. Nations must be induced to participate in formulating the kinds of policies, we wish for a New World Economic Order, not forced to simply stand at the back door and wait for somebody to hand it out to them as a finished product.

So I say, despite the difficulties, despite the problems of feasibilities as I've indicated, the problem is a soluble one. *We have the knowledge and means to solve the problem.* We face the difficulties, the political and diplomatic difficulties, of coming to an agreed form of solution in detail, to a solution in principle. These should be readily available. People may ridicule us and say, "Why are you doing that? You have no assurance that that will ever come about." And our answer is, "It's the only thing worth trying to bring about!"

# Great Infrastructure Development Projects: the Case of NAWAPA

I am delighted that this [film on NAWAPA] was made available and that I could have it shown today. It is slightly dated, as you noticed. The Great Lakes are up again, and they were very low when that film was made. But, other than a few variations of that sort, all the data are factual. The price levels are up slightly, but so are the income levels. Basically, that is a sound plan that we have considered over a period of time.

As was said, I served in the Senate for 18 years, and I have been concerned with the need to protect, conserve, and utilize our planet's marvelous supply of pure water. The absolute *sine qua non* is water. In a speech on the Senate floor, Senator Bob Kerr of Oklahoma once said that the time would come when a barrel of pure water would be of more value than a barrel of oil. Imagine that from a tycoon like Bob Kerr. Twenty-five years ago, Senator Kerr foresaw an expanding population

Senator Frank Moss *served in the U.S. Senate as a Democrat from Utah from 1958 to 1976, and chaired subcommittees of the Senate Interior and Public Works Committees. In the mid-1960s, he championed the cause of the North American Water and Power Alliance, NAWAPA, a continental water resources development plan developed in the early 1960s by the Ralph Parsons Company of California (see appendix). Before Senator Moss's speech, a short documentary film on NAWAPA was shown.*

and the continuing degradation and loss of our supply of pure water. The world could be on a collision course with disaster if we do not take steps early to carry out something akin to the NAWAPA plan.

Is it possible that any substance as universal as water could be in such short supply as to threaten famine and disease? The sad answer is "yes." A vast supply of this renewable resource is stored in icecaps or flowing into salty oceans in many places while enormous areas of our planet are desert, dry and inhospitable to life, either human or animal.

My own state of Utah has turned much of our landscape into beautiful homes and farms by diverting water from our mountains to our barren valleys. We started 100 years ago, but we still need more fresh water.

My political efforts within my party, and in the Senate, have pressed for water conservation, diversion, and use. We have a proud record in this effort, but it is not enough. We must raise our eyes, and expand our scope. My advice is: Make no little plans.

The world population now has exceeded 5 billion people, and with demographic projections of 10 billion people by the turn of the century. Plagued already with water shortages, spreading water pollution, desertification of vast areas of our planet—the only home on which *homo sapiens* can rely—and with world hunger and overcrowding confronting us in several areas, we must do something while there is time.

Of course, one course of action is to limit procreation. This has been done in some areas with limited, small success. Another is to improve food production with better seed, fertilizer, and pest control. But this, too, makes for only limited relief in selected areas. A third way is to abandon humanitarian and scientific efforts, to terminate medical efforts to fight disease, and thus we could invite back the diseases which in past ages swept away our children and whole areas of our population. Thus we might keep our population around 3 billion people. Of course, no sane person could subscribe to this solution.

The only reasonable and humane planning and action for the next decade is to conquer our desert wasteland, expand our areas of habitation, and expand our food production. This we know how to do. Vast areas of our planet now barren and desolate will become habitable and productive when we add water.

There have been proposals on our continent, proposals on the African continent, and proposals on the Indian subcontinent with the same expansive area. Nicholas Benton, who works with you, has worked in that area, and has published some very interesting, instructive data on it.

We must conquer this drought, the desert, hunger, overcrowding, and despair of our fellow human beings on this planet. Our time is running out. We should begin at once. When a person sets forth ideas and plans for water conservation and use, I give my support.

## We Have the Technology

NAWAPA is one of the proposals which I held public hearings on in 1965. But no concrete action was taken. International vision was too limited. Later, however, the Canadians did build part of this overall plan with the James Bay project, impounding and using water flowing toward the saline Hudson's Bay. International cooperation must come to this gigantic proposal if we are to prolong, and make better, human life on this earth.

As the film said, and as I believe I referred to earlier, this does not call for any new breakthrough. This is all technically feasible, within the engineering and technical capabilities which we have. So we are not asking to do something that is still to be developed. We may make developments along the way. One suggestion is that we might even find a way where we could use nuclear power if we could control and contain any of the radiation that might come. But that would be a bonus if it came along. We can do it simply with what we have in hand now.

If you think of what a change that would make immediately, to start to build what is called the infrastructure, to which there have been references made in this conference, on this

# Senator Frank Moss

continent, and we begin at once using materials and machines, and employing people, it would make an enormous change.

The other thing is that it would be a continent-wide development, with three countries involved in it—Mexico, Canada, and the United States—with benefits to all. One of the problems that we ran into when we were holding hearings and trying to get legislation started on this before, was the parochial interests of the Canadians. For some reason, the ones who were then in power opposed the plan. By the way, that government has changed and it is much different now, I think. I was invited up to speak before the Royal Society of Canada on this plan. General McNaughton, who then headed the Canadian ministry that oversaw its water, was on the same program, and it turned out to be pretty much of a debate. It boiled down to the fact that Canada was distrustful, that some way or other they were going to be leaned upon, or taken over, or dominated, by their big neighbor. I don't think we ever overcame that.

But it need not be that way, because Canada would benefit as much as we, and, anyway, the source of the water would be reimbursed. We pay now for water when it is used. Water is a commodity. And it is a renewable resource. So the area from which the water would be diverted would be reimbursed for whatever the value of that water is, and it would go on, year after year. I tried to say to General McNaughton, "You are down trying to sell us wheat, oil, and minerals. You want to sell them to us. Why don't you want to sell us water? Once a mineral is delivered, it is utilized and gone. But your water will be there next year. And it will be there the year after. And it will be there the year after."

The other thing I would point out is that the Parsons Plan, the one we were looking at, would use only, at maximum, 20 percent of the flow of those rivers. It isn't going to dry them up. It takes some of the water, and it would be taken at a place where it is already in the frozen, northern zone. I went up to take a look at the mouth of the McKenzie River, and there's nothing up there but muskrats. Yet the McKenzie dumps 264 million acre feet of water directly into the Arctic Ocean every year. So if you took 20 percent of that and sent

39

it southward, the muskrats wouldn't even be bothered. They'd still be there. There's nothing else around, really. That is the same with the Yukon River, as it gets out to where it dumps into the ocean.

So it is not as though you were going to change a whole landscape up there. You are only going to take part of surplus water. That's the other thing I want to underline: surplus water, which means that if there is any valuable use of water, those people who are there now can expect to have that beneficial use of the water. It's only when you've got surplus. The other thing I said to them is, "You ought to do your homework, and make a survey and decide exactly what water you project that you'll need in 50 years or 100 years. How much do you need? Then add that and see if we've got any surplus water here. If there's surplus water, we ought to bring it and use it."

We can then take this surplus and bring it on down to the northern part of Mexico, where the land is very fertile and the crops grow very well, if they can get water. It always comes down to that.

I have long since left the Senate, and I have no economic axe to grind on this. But I am so convinced of the need for this that whenever I get a chance to say something about it, I do so. I accepted the invitation to come here today, and I'm glad that I did because I can talk to people from all different places, even people not from this continent—leaders—to see if we can't keep alive this idea of doing something basic and needed, such as this transfer of water and utilization of it to the betterment of our whole landscape.

# The Brazilian Debt Situation

Mr. President, Mr. and Mrs. LaRouche, Ladies and Gentlemen:

I don't speak English. I will speak in Portañol, a mixture of Spanish and Portuguese. But, I will speak with my whole heart. And I tell you that I have left my country in these days in which Brazil writes its new Constitution after 20 years of dictatorship, and why I believed that here, in this meeting in the United States, I would be able to find people, from wherever they may be, who think like we do, that it is not possible for some men, for some peoples, to live well, with the heads of men from other regions under their feet. That it is not possible for men to be happy knowing that, what they have too much of, is what is needed in other countries to kill the hunger of the children who die of hunger.

And, in Brazil today, because the debt is too big, because the international banks force us to pay what we do not have, there are so many children who die of hunger every day. Some days ago, speaking with the governor of the state of Pernambuco, he told me that in his state, his province, in some areas, of every thousand children born, 500 die before their first birthday, that is, half of all of them die.

UNICEF told us a few days ago, through the international

Irajá Rodrigues *is a federal deputy from Brazil, a member of the Foreign Debt Commission, Brazilian House of Deputies; and president of the Latin American Organization of Municipalities.*

41

press, from near here in New York, that last year in Brazil, 60,000 children died, on account of the debt. We are paying too high a price. And it is said, "No, Brazil suspended payments and made its moratorium last year." No, we stopped paying only $4 billion, but we paid $8.6 billion in debt service. I am a member of the Foreign Debt Commission of the House of Deputies. And I will be, a few days from now, president of the Latin American Organization of Municipalities. I repeat, I expect to find here men and women who think as I do. That's why I'm here.

I am going to tell you that we in Brazil know what we owe—$120, $130 billion. And, if we added in the internal debt, we would certainly count $200 billion. But we Brazilians do not know why this debt was contracted, why it was built up during the period of the dictatorship. Anyone who said something that was not favorable to the government could only say it once. We don't know correctly if the $120 billion was money, was resources, was something that effectively reached Brazil. Most is the product of the never-ending payment of interest, commissions, and capital. One part could very well be represented by [commissions to] intermediaries. That's true. The people of my country received very little, much less than what is now being demanded of that very people.

Therefore, it must be proclaimed everywhere, all over the world, in the loudest voice, that Brazil cannot continue to pay its debts with the blood, with the death, of its people. And, therefore, we presented the National Constituent Assembly with a proposal that its final disposition [Constitution] include an article which states that Brazil will not pay anything of the foreign debt for the coming five years. And that it is going to verify, with an audit, what the true debt is, how much the country really owes, because things cannot continue this way.

Right now, between continuous negotiations, ministers from my country are returning to the United States to try to convince the international banks to give us a few dollars more, merely to roll over this debt. And then it is demanded that Brazil go on its knees before the all-powerful International Monetary Fund. And, lower than on its knees, that it spread [its legs] and that it seek to pay ever higher interest rates through recessionary economic policies. That it also seek to reduce the consumption of a population that earns very little, and for that reason, lives very little—a population in my country whose workers have a minimum wage of $40. I say $40 not for a day, nor for a week, but $40 a month. Even if it sounds incredible, the international banks are saying that it is necessary to further reduce the conditions of access to the most indispensable things by those who are not unemployed, but are the workers of my country.

And it is said that we have to obtain a $12 to $15 billion [trade] surplus per year to be able to come up to date on our international commitments. And, at the same time, reductions are being made in every way, through exceptional rates, the reduction of our conditions of exploitation.

If we do not have the right to sell what we produce—our shoes, our airplanes, our microcomputers—how do they expect us to also have favorable surpluses on our trade balance to be able to place the dollars in the hands of our international creditors? Doubtless, what they want is to squeeze the juice—the blood—from these men. We cannot, and we will not, consent or allow this to happen. Brazil's honor requires that this not be permitted to go on.

But our situation is no different from that of the other Latin American countries. It is easy to see what they did to Mexico. It is easy to see what they did to Argentina. You can see what they are trying to do to Peru, to Ecuador, and to all the Latin American countries.

It is necessary to stop this situation.

It is not enough that those men who live well were happy to know they live well because for each of them there are thousands of children who are hungry every day, who, thanks to them, die a little of hunger every day.

## Latin American Unity

Thus, I tell you it is necessary to unify Latin America. The final declaration of a recent conference in Brazil, attended by eighteen Latin American countries, representatives of the parliaments of those countries, representatives of the international parliament's Latin American section, left it quite clear that, were this situation to continue, Latin America, as a whole, should immediately stop paying capital, interest, commissions—*all* Latin America's foreign debt, understanding that that is the only solution.

They did not leave any other door open for us.

A few days ago, I was here in the United States on a mission with a few other congressmen. We were told clearly that, in the coming fifteen years, the international banks would not give a single dollar more to Latin America. And, on returning to Brazil, I asked the President of my country if anybody knew that the bankers would not give a penny, a dollar, a cruzado.

What is this? We know the time has come for the men of Latin America to unite, and more than the men of Latin America, the men of all the world who think that only through the development of all peoples is it possible that all men live well. The time has come to declare that we want to build a new world upon a new social reality, and to reach that, it is first indispensable to build a New World Economic Order, without the bitterness of having to see some living well, while others live amid the stench of death.

That is why I am here, to say that we from Brazil want the building of a new society, a New Economic Order, and if that is what unites you here today, and it seems to me that is your will, I say: Count on us, on me, on those in Brazil who think as I do, on those in Latin America who are tired of seeing our blood let, of seeing our lives, our blood, our gold, the fruit of our labor merely fatten the pockets of the international banks.

# The Effect of Debt Pressure on Perspectives for African Trade

I have to present three announcements before I start my speech. I am representing the African Association of Trade Promotion, which is a group of twenty-four member states in Africa, and is an intergovernmental institution. My second announcement is to apologize for the jet-lag which I got in traveling. The third announcement: my tongue is Arabic and not English, so I must apologize beforehand for not good English.

Mr. Chairman, this is a great honor for me to speak before this gathering on some aspects of one of the difficult problems facing the world now, and most probably for decades to come. This is, of course, the African economic crisis. I will confine my speech to one special, narrow area covering African trade, as the weakest point in the whole structure. And also the lowest priority sector in African planned activities.

The logical start of my speech is to remind you and probably brief you about the existing economic situation and trade relations of the African countries. First, it is well known that

Farouk Shakweer, Ph.D., *of Egyptian nationality, is the secretary general of the African Association for the Promotion of Trade, headquartered in Tangiers, Morocco. The AAPT, representing some twenty-four African countries, is associated with the Organization of African Unity (OAU).*

Africans are the poorest people on our globe. But some of us may not know that we are, in Africa, going to be poorer and poorer. And no such indication that will change this fact can be seen in the future. Second, after being hit by drought for nearly five consecutive years, the situation became unbearable. About 5 million Africans are dying every year of this drought.

The United Nations' special declaration on economic crisis in Africa has accepted the development of special emergency measures to help Africa through special, international, combined efforts. I have to indicate that in this session of the United Nations, the radical solution presented to help Africa is not really fair enough to cope with the existing inhuman situation on the continent. However, the program accepted by the United Nations takes the name APEER, which is an abbreviation for African Priority Economic Emergency Program, and it is directing its attention on two main symptoms. The first is to activate and rehabilitate production in Africa, to cope with the problem of hunger facing the majority of the continent. The second priority sector is to develop industry in combination with infrastructure, to build up the necessary link between agriculture and food production on the one hand, and other sectors on the continent on the other hand.

## Africa's Untapped Potentials

Existing data indicates that Africa is in the middle of a battle for survival and development. Nevertheless, it was also clear that in spite of the human and social tragedies of the immediate past, and the economic retrogression over the last few years, Africa remains a continent of great and tremendous potentials, almost all of which have so far remained untapped.

African states are now most anxious that those potentialities be realized, so that their continent should not continue to be the weakest link in the network of economic integration. Directing my attention to African trade, which is the main topic, I am going to cover external trade of the continent with non-

African countries, and also intra-African trade, which deals with trade among the African states.

First of all, trade is the weakest sector in the whole structure of the continent. And because of these weaknesses, debt accumulated rapidly and reached over $200 billion by 1987. The debt service amounted to more than 50 to 60 percent of the export revenues of the continent. And it is well known that almost 20 countries in Africa have now started to discuss the probability of stopping payment of their debt service. In spite of its weaknesses, the sector supplies more than half of the revenues of most African countries, although it is very weak, but still supplying the governments of Africa with 50 percent of their revenues to manage their administration and government activities.

Trade is one of the crucial instruments to improve economic performance, and is an essential element in the implementation of the latest plan of action. I have to indicate here that the latest plan of action was declared in the year 1980. One of the major targets for the 50 countries of Africa was to develop the common African market in hopes of stopping deterioration of the African competitive situation among the other countries of the world.

Ninety-five percent of African exports are primary commodities, and these commodities are in the form of raw materials and minerals. This type of export creates difficulties, in particular in the form of stagnation for the economy, fluctuation in export earnings, instability of commodity prices, and impact of the collapse of such prices on resources, which also prevents any solid economic forecasting for the development of Africa's future.

The most persistent and striking feature of the foreign trade of Africa, is that it is overwhelmingly with countries outside of Africa. Ninety-six percent of foreign trade is outside of Africa. The share of African international trade in world trade is declining year after year, and is now on the level of 4 percent of international trade. A few years ago, it was more than this. For example, Africa's share of foreign trade has fallen from 4.2 percent in 1980 to 3.0 percent in 1985, an overall loss of

47

ground of 25.6 percent, which means that we are losing day by day.

Intra-Africa trade, on the other hand, accounts for only 3.9 percent of the international African trade, i.e., African nations are trading only very, very minimal amounts with their neighbors. And this is because of many reasons, basically historical, colonial reasons.

Another feature revealed by analyzing the data for the years 1980-85 is that foreign trade registered an annual negative rate of change of $-.3$ to $-.4$ percent for these years, but African trade changed, more rapidly at $-6.05$ percent. Not only that; overall, while African imports declined by $11.5 billion in this period, the export of African states declined particularly in this time, by as much as $54.4 billion, as a result of which, Africa was a net exporter in 1980. In 1980, it became a net importer for the following year, and remained so for the rest of the period. . . .

Africa's major trading partners are not African. Led by the developed market economy countries, for historical reasons, the European countries are still the main partners, taking almost 50 percent of our trade. The United States is in second place, and the situation is not very much different from either the export or import side.

Africans generally export primary commodities in exchange for mostly manufactures from other continents, both processed and non-processed items. The items traded by Africa are mostly the usual international trade composition: food items, agricultural raw materials, fruit, fertilizer, minerals and related material, chemical products, iron and steel, non-ferrous metals, machinery, and transport equipment.

## Africa's Lack of Infrastructure

The main features governing African trade are as follows:
(a) Africa consistently spent more on its food imports than it received for its food exports;
(b) this pattern was repeated, because of the backwardness and underdeveloped situation on the continent;

(c) the African states registered a positive balance from exporting agricultural materials, fruit, fertilizers, minerals and non-ferrous metals. However, the negative balances deriving from (a) and (b) above, were so large that they led to an overall negative balance of African trade.

Such details of African trade should make clear that change is very important in the near future. The requested efforts are really what we currently have indication of from the balance of African trade and the composition of African trade.

The existing weaknesses of African trade are due mainly to exogenous and indigenous factors. The indigenous factors are basically due to the following:

(a) inadequate basic infrastructure: roads, waterways, railways, are almost nonexistent. This type of facility is basically required to transfer goods from producing units to consumers;

(b) the trade customs and foreign exchange regulations are established and built to favor imports from developed market economies and to export raw materials, and not in favor of developing African trade among ourselves, or giving incentives for the new industrial development of the economy;

(c) in Africa itself, the banking system and financial houses dealing with trade are mainly dependent on the backs of institutions of the West, i.e., of the financial circles. It is difficult to accept a letter of credit from another African exporter or importer, without getting a confirmation from a European bank.

Morever, the situation is getting worse, under the existing economic crisis of the continent. So, the situation is relatively dark, day after day. Many countries are now on the edge of stopping payment on their debt service, and curtailing their bank loans to the lowest minimum. No doubt, a situation of this kind is going to develop unfavorably and violently, not only in Africa, but all over the world.

So, proposed solutions should be presented. The proposed solution, as a matter of fact, is not very much different from the insight in the inspired speech of Mr. LaRouche. It could work along the same lines. I have to indicate that now in Africa, we are in the middle of assessing the existing financial situation and its fitness for the continent.

# Development is the Name for Peace

For over 25 years, since African independence, and the establishment of the Organization of African Unity and the African Development Bank, the established financial system has not conduced the requested results, and Africa is still the poorest, and is going to remain the poorest.

What they call "project financing" means that you have to confine yourself to a specific feasibility study, going to a specific place, working out the layout of the place, looking for the skilled professionals, and this takes a lot of time.

Last week I was in Khartoum, discussing this financial situation with the African-Arab Development Bank, established by Arab countries—mainly oil-producing countries—to help Africa. Every year this bank spends almost $200 billion in Africa. This bank has started to realize the same situation as the African Development Bank—the project development time is too long.

You stay two to three years to supervise the financing of the project. Then you stay another two to three, or five years to establish the project, and then, after two to three more years—they call it the "trial and error" stage—what do you get? A new generation comes in. Many, many variables in the economics of the continent change, and even the value of the project in this very quickly changing world, is going to have no sense at all.

So, I am in the process of establishing a new system for Africa, to finance the trade, which is very effective. Most of the weaknesses in Africa come from the trade side. Now we are establishing an export-import bank for the continent. This is the first time in history that we are establishing a bank to deal with exports and imports for 15 member states.

What will it look like? How will it be financed? We are looking for big brothers like the U.S. Export-Import Bank, the Japanese Export-Import Bank, the Korean Export-Import Bank. How are we developing assistance? We are working our best, but I have to ask also for the support of the people who are looking for a new economic order, which we are really depending upon.

# Who Is Responsible for the Coming Crash?

Ladies and gentlemen, I would like to take this opportunity, very quickly, at the beginning of our afternoon panel, to situate us in the ongoing process of the economic crash, and also to name the names of some of those who are guilty of bringing humanity into its present plight.

We are living in the middle of a collapse of a financial bubble. This is nothing new in history. It is similar to the collapse of the Bardi-Peruzzi financial bubble back in the 1200s. It is similar to the Spanish and Portuguese bankruptcies of the 1500s that wiped out the Fugger banking empire, centered in Augsburg, Germany. It's similar to the tulip bubble, the Mississippi bubble, the South-Sea bubble of the eighteenth century, the panic of 1837, the crisis of 1929. Back in the fourth quarter of 1987, we had in effect a financial earthquake that reached the level of approximately seven on a "financial Richter scale." We now expect that during the course of the first quarter of 1988, it is very likely that that earthquake will be repeated at a level of perhaps seven or eight on that same "financial Richter scale." It is then exceedingly likely that within the second or third quarter of this year, we will have a more definitive, cataclysmic earthquake which will reach 10 or perhaps even 11 on that same Richter scale, and of course,

Webster Tarpley *is president of the Schiller Institute.*

it's a logarithmic scale, so they get very big, very fast as you go up the scale.

This has meant, of course, a stock market crisis which has proceeded around the world to wipe out one-fourth, one-third, one-half of the values of most world stock exchanges.

At the same time, we have had a catastrophe centered upon the United States, which has been the rapid decline in the value of the dollar. We are now heading for a dollar that may be worth 1.20 [German] marks, 100 yen, or even less. One of the symptoms of the prevalent insanity in Washington is that Washington officials are not even sure that this is a bad thing. Even the rather imbecilic oligarchs of the British empire in 1949, when they had to bring down the pound from about $4 to about $2.80—even they knew that this was a catastrophe for them, something much to be avoided.

It of course means that the viability of U.S. Treasury bills, notes, and bonds, is under question; and it is no longer certain how long the patience of Japanese, European, and other investors will last, leading them to buy these kinds of paper.

At the same time, as we're going to be hearing about during this afternoon panel, we have the phenomenon of the implosion or collapse of whole national sectors: Large countries in the developing sector, above all Mexico, but certainly other countries in Ibero-America, who are undergoing combinations of hyperinflation and economic collapse.

The perspective for the immediate future is that the panic that we've become familiar with in the stock market will begin to kick over into real estate values. It's perhaps enough if I mention names like Helmsley-Spear of New York City, or the picaresque figure of Donald Trump, the highly leveraged super-landlord of Manhattan, to give you an idea of the kinds of highly leveraged, highly inflated real estate values that are now in the process of collapsing. The combination that this introduces is that, since so much of the resources and assets of the banking system are held in mortgages, the banks are going to be caught in a situation of defaulting payments from debtors at home and abroad, then complicated by the collapse of real estate values and the dubious quality of mortgage paper.

Webster Tarpley

# Local Banks Threatened

This then leads rather rapidly into a situation of the type we had in 1932, when 2,000 banks went bankrupt. In this past year, it was only about 200, but we're now getting back to the kind of "ballpark [figure]" that we had in the 1930s. It's one thing to see the money-center banks go down; the Bank of America is of course bankrupt; the principal New York City banks are bankrupt; Continental Illinois has gone bankrupt once, and it's now having the rare privilege of going bankrupt twice, within the period of a decade. But what is more important, is the banker on Main Street, the banker in the home town, who provides payrolls, who discounts commercial paper, accepts bills of lading, etc. Were those local bankers to be shut down, then all the wheels of the economy would grind to a halt.

It is impossible to run a capitalistic economy without a local-regional banking system. It means, among other things, that if those banks were to be shut down, you could have a situation of tens of millions of people thrown out of work in a very short number of weeks; and you might have a situation where the lucky ones, the people who still have some money to spend, could go to the stores and find nothing to buy, because distribution and transportation cannot be conducted without a functioning credit system. At that point, we are confronted with unspeakable conditions, beyond the imagination of most of those now observing the collapse of this economy.

In that connection, the kinds of emergency powers that Mr. LaRouche talked about, which he would do as President, can be roughly compared to the Roosevelt bank holiday of 1933. Roosevelt had a very crude option of shutting down all the banks for a month to then reopen them, by keeping them solvent and reorganizing them over a period of four weeks. Today, with the Federal Emergency Management Act, the Defense Production Act, and other kinds of legislation, the President has the ability to do that in a much quicker and more fine-tuned way.

DEVELOPMENT IS THE NAME FOR PEACE

As the crash goes on, the kinds of luxuriant vegetation that we've seen in particular over the past five years—people like Ivan Boesky, Carl Icahn, T. Boone Pickens, Edmond Safra, the Gordon Gekkos—on Wall Street, are being swept away. These are the kinds of people who have been conducting off-balance-sheet lending, asset stripping, highly leveraged junk bond-assisted hostile takeovers of the Drexel-Burnham-Lambert variety, and generally wrecking this economy. The carnival for these parasites is over, and Ash Wednesday has definitely arrived.

## A Look Back at Bretton Woods

At this point, let us just take one look back at Bretton Woods 1944. What was the essence of this system? Whose interests did it serve? Who is therefore responsible for the kinds of things that are going on today?

The Bretton Woods system was not devised to help the system of sovereign states, but rather served the family fortunes of a very limited group of oligarchs and aristocrats. In particular, at the center of the Bretton Woods system were the so-called *fondi* or *stiftungen*, the family fortunes of the European feudal aristocracy. These are people we know here in the U.S.A. with names like Astor, Morgan, Mellon, Lowell, Roosevelt, and their European titled oligarchical counterparts.

Let us remember that, at Bretton Woods in 1944, the leading player on the part of the U.S. government was Harry Dexter White, a Russian-origin individual, the son of recent immigrants from Russia, who was later on accused of being a Soviet agent. He was working for the secretary of the treasury, Henry Morganthau, the same man who was author of the Morganthau plan.

The presiding demi-urge of the Bretton Woods conference was really [former Nazi Finance Minister] Hjalmar Schacht. The purpose of the Bretton Woods system, as you can see through the influence of Morganthau, was to extend to the entire world the kind of looting system that Schacht had been able to create, primarily in Central Europe. It was a system based on rigged parities, rigged in terms of trade, giving the dollar the ability to loot and drain the resources of Latin America, Europe, and other parts of the world. It guaranteed that there would be no development in the Third World, and it guaranteed the absolute emiseration of the developing sector.

This system then crashed down between 1967 and 1971, with a kind of postlude up until 1973. The people who ran these operations were the central bankers, servants of the oligarchical families. It's enough to think of the Bank of International Settlements in Basel, run by Fritz Leutwiler, of Swiss finance. Think of Guido Carli, the head of the Italian Central Bank, who invented the Eurodollar market, later on, the Eurobond market, and now the Euro-junk bond market. Then we have Paul Volcker, another distinguished servant of the oligarchical fondi, who was able to bring in the innovation of the highest interest rates in the Christian era known since the time of our Lord. And others, like Emminger, Perle, William McChesney Martin, Robert McNamara, Barber Conable, Jacques de la Rosiere, Camdessus, Pierre Paul Schweitzer, and others. I think the point is clear. These are the people responsible for the present situation.

## The Bretton Woods Ideology

The other aspect of these matters is the ideology that made this possible, an ideology promoted at each point by these same fondi, in particular by the Venetian fondi. This goes back to ideological operations of the 1950s and 1960s, manipulations of the world spirit promoted by the Societé European de Culture of Venice, the Cini Foundation, and later, the group that they spawned, the Club of Rome of Aurelio Peccei, Alexander King, and others.

These ideological manipulations came upon the American scene in the form of such things as the "Triple Revolution Committee," which sent a letter to President Johnson back in the mid-1960s recommending that the United States cease to be an industrial society, but rather become a service economy and a consumer society. This is indeed the course that Johnson ruthlessly pursued. It is the notion of a "post-industrial" society, or a "technetronic" society, as described by Zbigniew Brzezinski, which is really the cause of the situation of the world and national economies at the present time. This is the argument that there are "sunrise" industries—video games, pizza delivery, and hamburger sales—and "sunset" industries, primarily the backbone of commodity production, the "smokestack" industries, as they were called. This ideology has promoted what is now virtually the total cessation of commodity production inside the United States, in agriculture and industry, as these areas grind to a halt.

We had meant to have our conference in Bretton Woods, up in the White Mountains, but we don't have the drama of that symbolism. There nevertheless is one thing that we have here in Andover, Massachusetts. George Bush is making his bid to become President of the United States. His friend James Baker is running the Treasury, primarily as an adjunct of that campaign. Not so far away, across town here in Andover, we have the place where George Bush became a "preppie." That is the Phillips Academy, not likely to become the place that people will make pilgrimages to in the near future. If you go and visit George Bush's room, over on the Andover campus, you can see that there were bars placed in the window by the headmaster at the time!

Now, George Bush and James Baker are not conducting a long-term solution to this crisis. There are others who are thinking in a somewhat longer-term frame of mind. For example, the Morgan group, J.P. Morgan, Morgan Guaranty, Morgan-Grenfell, on Wall Street, who are promoting things like the Bipartisan Economic Committee, with Felix Rohatyn and other investment bankers. One of the things that they have decided to target is the Social Security system. The average Social Security check is now $400; that means that half of the Social Security checks are less than $400; there are now approximately 37 million people who owe their existence to these minute Social Security checks. These are the senior citizens, 30 million and more, living, or dying gradually, on dog food, cat food, in a situation of absolute economic misery.

All the leading Republican candidates, Bush, Dole, and the seven dwarfs on the Democratic side, are all committed, each in his own way, to cutting, looting, and sacking this Social Security system.

Then, there's an even longer-term solution being promoted by the Venetian financier Carlo De Benedetti, who thinks that by the end of this century there will be between ten and twenty banks left in the world, and he plans to dominate them. He is the leading operative of the Assicurazioni Generali Venezia, the largest insurance company of the world, and the largest administration of family fondi of the European feudal aristocracy. De Benedetti is now in the process of trying to take over an organization called the Société Générale of Belgium, which is a holding company that dominates the entire strategic raw-materials production of Africa, in such areas as Zaire and the former Belgian colonial empire. De Benedetti's goal is to create a European-wide holding company, that would then tend to dominate the post-depression world. The policy would then be a Schachtian economic policy in an extreme form.

The only alternative that is posed to these kinds of reorganization schemes coming from the Morgans, De Benedettis, the Rohatyns and others, is a recovery program: The program which accepts the reality that we are now in a world depression, that there is a financial crash ongoing, and then provides an economic recovery as the only means of combatting that depression. It is only by supporting such an economic recovery program as Mr. LaRouche has outlined it, that we can save ourselves from the clutches of these feudal oligarchical families, who have been the managers of the world economic system since 1944, and who are responsible for bringing humanity into its present condition of agony.

# Emergency Actions to Stop the Depression

I was asked to address the ideas that some people are using today to deal with this economic collapse as it occurs.

Oklahoma has a mass of about 150 miles by 300 miles and about 3 million people. So that makes us a nation about the size of some of the nations of people that are here talking. Also, we are almost like a Third World country, as a manufacturer of agriculture products, like most of you. And we're also a heavy producer of minerals—oil and natural gas. And we're probably treated like a Third World country, just like anybody else.

You know, as we look at what we can do in the states, you have to realize that the states have limited abilities. We don't have the right to print money, and the collapse is going to be manipulated somewhat around the control of money and making it available. However, there are a number of things that we intend to do. There is legislation that we're drafting now, trying to have it ready in case a collapse occurs, and most of us believe that it will.

If it does happen, there are certain things we can do. We need to implement some things to keep our society moving and not have a total devastation. We're not allowed to print

*This speech was given by a member of the state legislature of Oklahoma.*

money. States don't have the right to do that. However, we do have the right to do certain things. So, first of all, we've got to put a control on the closures of banks. As you are very aware, you've got to have the money supply. We can't let those banks be taken over or go under. We've got to put stops on the foreclosure of our money systems. Second, we've got to put the stop on loan foreclosures, because people won't have the money to pay their notes. The homes, the small businesses, the private enterprises, the farms, the oil industries that are very abundant in Oklahoma—these things have got to be prevented from being taken over. This requires a moratorium. It will also require a moratorium on interest during this time period.

Another thing that we've got to do is to put some liquid capital into existence, which is difficult when you cannot print money. The thing you can do is to issue vouchers—to people in the oil industry, the agriculture industry, the food manufacturers—the state has the ability to issue vouchers for those things. They're not legal currency, except that in the State of Oklahoma, they can be used to pay state taxes. And industry, other businesses, can take these things, as many as they want to. They aren't legal currency, but they can be used for exchange. You see how that works?

We've got to put these things in motion just to keep our vital system from going under.

[Another] thing that we want to do immediately is to change the educational system. We've got an educational system down there that is one of the most expensive things that we've got, yet only 75 percent of the graduates from the program are literate. That means that 25 percent are very semi-literate. They're almost illiterate. . . .

The [next] thing that we want to do is to terminate many of the exclusive franchises that are going on in the State of Oklahoma. In Oklahoma, we have to buy certified seed that has come out of our own universities, developed there. Now it's controlled by five companies and they sell these things— you're not supposed to even take seeds off of them and replant them. These franchises can be terminated and that needs to be done.

Also, there's a number of industries, like cattle, grains, oil products. A lot of these things are controlled by a very small group of people. We've allowed them to be put into monopoly control. I think the states have the authority to overturn these monopoly controls. And we need to just push in the legislation to make that happen.

However, I want you to know that the states cannot take care of this problem in the long run. That's got to be done with the national government. That was set up in our Constitution, and we have to go back to our own U.S. Constitution. Our U.S. government needs to issue its own money at no interest to itself and put it into existence by loans to people. And when it does that, the taxes will go down considerably.

## Go Back to Barter

. . .I think that we ought to allow international barter. If we've got something in Oklahoma like pecans or carbon black that Egypt wants, or some other country, we should be able to barter that—for bananas—or whatever we want to do as a state, and get away from this international trade thing. The international trade people are skimming off of everybody, and we're the losers. I've been in international trade, and the only ones who were making anything were the ones who were handling the money. So we need to change national laws to allow local governments to trade freely. And that can be somewhat of a barter system.

We need to change our election laws to make it so that national and international power brokers don't come into local legislative races and buy those legislators or come into a district and buy a U.S. congressman. They're bought and paid for, most of them. A lot of them before they ever get into a position. It's very difficult to be a legislator in that situation. The election laws have got to be changed. The monopoly laws have got to be changed.

I have a copy here of some of Oklahoma's emergency legislation. Mostly what this is, is an education system, because legislators are as ignorant as most people out in the community. It's important that those legislators be educated, and that's what we're trying to do.

I do agree with the solutions that are coming out of the conference. Thank you very much.

# The New World Economic Order and Brazilian Debt

Mr. President, Mr. and Mrs. LaRouche, participants:

I am here representing Deputy Eduardo Lima Filho, of the Brazilian Democratic Movement Party. I am from the University of Brasilia, and have worked over the years with the members of the party who worked for a long time to bring down Brazil's military dictatorship. Deputy Eduardo Lima Filho was one of the fighters, as was Deputy Irajá Rodrigues, who spoke earlier.

I would like to say that we have assembled to speak to the question of the New World Economic Order. When we analyze questions on the Third World's foreign debt, we reach the conclusion that the international banks, institutions like the International Monetary Fund, and the behavior of the governments of the advanced countries, have been causing a real international economic disorder, a complete disorganization, especially of the debtor countries.

The Brazilian case amply demonstrates that the creditors lack the political will to find a solution to the debt problem. The Brazilian case is one in which the creditors are running

Professor Dercio Munhoz *is an economic advisor to the PMDB, the Brazilian Democratic Movement Party, the ruling party of Brazil.*

no risk of losing their money, because the debt is a debt owed by companies, by government companies or by private companies, by companies which are in very good condition, by companies which pay interest and principal on the debt.

But, how can we transfer debt interest or principal in hard currencies? If we could print dollars, we would be able to pay off the debt. We could pay interest, if there were conditions permitting trade. Therefore, the companies are able to pay interest, and they include that interest in the prices of their merchandise. Thus, the population pays that interest. This is what Deputy Irajá has described, which causes great difficulties to the population, especially the worker population.

If the creditor countries allowed us proper trade conditions, we would be able to promote increased production to have merchandise to export. We expect countries like Brazil, which have very diversified exports, to find support for new investments to produce goods for export. In this case, the creditor countries would, at least, be trying to solve the debt problem.

But what really happens? In reality, the hypothesis posed by the creditor nations, the international banks, the International Monetary Fund, and the World Bank, is that we do not have [spare] productive capacity and that countries like Brazil are under conditions of full employment. Another part of the hypothesis is that we are having troubles because of an excess of internal consumption, because the population has very high consumption levels, and for that reason we have trouble paying the debt.

So, what do they do? The political solution recommended— really imposed—by the International Monetary Fund, with support from the creditor banks and countries, is to artificially create surpluses for export. We could produce more. But no, the path is to create artificial surpluses. How is this done? Increasing the domestic interest rates charged by banks, increasing taxes, reducing family incomes in order to reduce consumption demand. The worker families of the debtor countries are forced into a classical process of impoverishment.

When the surpluses are created, incomes are reduced, products are cheapened. What happens? There are growing restrictions on trade, which is contradictory. Therefore, it can be seen that even the plans to create surpluses—which is inhuman—find barriers across which we cannot export.

The Brazilian case today has an additional barrier, which is that, since Brazil is trying to install a microcomputer industry—so that it not be totally dependent on foreign technology, so that we don't have our production and our trade under foreign control—we are under a daily threat of our exports suffering new restrictions in the American market.

The interesting thing is that the American government has not yet defined an additional restriction, but it is said every day that tomorrow, or next week, the new restrictions will come. The new tariffs which would be arbitrarily imposed, in violation of the GATT accords, would even be imposed, it is always said, even against past exports. Thus, Brazil has material ready for shipment which it cannot ship because the American buyers say, "How can I buy today if tomorrow the American government will make an additional charge?" What has happened? This is causing a situation of semi-paralysis of Brazilian trade—a purely political action.

Thus, we find ourselves in a recession after creating surpluses based on the growing impoverishment of our population. We also suffer from strangulation imposed by the Paris Club, the World Bank, and the IMF. How did this happen? Very simply, commercial banks want us to pay interest; and we have problems doing that. The Paris Club (which includes the U.S. Export-Import Bank, government banks from Germany, Japan and France, plus the IMF and the World Bank) wants Brazil to pay not only interest, but also amortize principal. That is, we would have to pay $10 to $12 billion in interest and the governments and international institutions want another $4 billion in amortization of the debts owed them.

What is going on? The country is being strangled. This is reflected in a fall of family incomes, a fall in consumption, and consequently, a fall in employment. The Brazilian economy is now semi-paralyzed.

The model imposed on the economy results in interest rate increases causing a major transfer of income from worker families to the private Brazilian banks. During the past two years alone, increases in interest rates in Brazil, put an additional $25 billion annual burden on the productive system. That is, Brazilian banks and their depositors increased their take by $25 billion; and worker family incomes had to relinquish [that amount].

Thus, the crisis is now deepening. This policy began in 1981 in Brazil, shortly before the debt payment crisis arrived. Brazil initially followed a policy like that the Monetary Fund recommends, without having signed a formal accord with the IMF. Before this policy began, the Brazilian productive system paid for internal and external interest appoximating 10 percent of the country's total income. Today, internal and external interest paid by the productive system has reached about 30 percent of national income.

That is, a third of the price of every product a Brazilian finds on the market consists of internal and external interest. You can see that Brazil is going through a very serious situation of impoverishment and disorganization. This is nothing new. I am not saying its situation is more dramatic than what we see in Mexico or Argentina. Yet, things cannot go on this way.

At the same time, what happens? Since Brazil has been made into one big casino of speculation, monetary policy—the abortion which the Fund wanted—the government's internal debt has also increased, and today that internal debt is about $6 billion. Also, increased interest costs brought with them increased production costs. Brazil will not be able to overcome inflation until interest rates are turned around.

## Democracy Threatened by Economic Disintegration

We went through two stabilization programs. In recent months, monthly inflation rates went from 6 to 8 to 12 percent, even 17 percent. This month we should reach 20 percent and possibly reach or pass 25 percent per month in February, a process of economic disorganization which sharpens social tension and thereby generates political instability. We, who took twenty years to get a military regime out of power, we, who tried to re-encounter democracy, now find ourselves faced with an economic disintegration process which threatens all our democratic conquests. Brazil needs to lower interest rates, and needs the income obtained by the banks to be returned to families, so that growth of production and employment can be restored.

We need fixed interest rates on the foreign debt. This is not something impossible for the banks. All that is needed is for countries also to have fixed rates on their international reserves deposited in the international private banks. We need lower spreads. Can you believe, Brazil paid $10 billion just for excessive interest rate spreads. Today, we have a debt a bit higher than $100 billion, but we have paid about $80 billion interest on that debt. As Deputy Irajá Rodrigues stated earlier, the origin of our debt is almost solely the accumulation of interest.

So, what happens? We in Brazil could have ended this decade with a [yearly] product worth $500 billion, but we are producing only $300 billion. We are paralyzed. Exactly the opposite is happening. We cannot produce, but they want us to pay the debt. They don't let us print dollars, but they want dollars. The international banks and the debtor countries do not want to print more for Brazil, but they want dollars. Where could we find those dollars? There is a political process of turning the affliction into political domination.

Given its productive structure, Brazil could export, but under these conditions, it cannot. Some countries are able to pay their debts with iron ore, or other primary products, for which there is a world market today, but will there be to-morrow? No! The debt question is insoluble, unless the cred-itors, in good faith, seek a real solution.

So, what will happen? Will Argentina disorganize itself, along with Mexico and Brazil, and not achieve a solution? The debtor countries have to have a priority. The priority must be to preserve their internal economic organization, preserve their internal political organization. That is the priority, be-cause if it were not done, then all their sacrifices would be for naught. Thank you.

MPINGA KALONGI

# The Necessity of the New World Economic Order for Zaire

When the leaders of the Schiller Institute in the Federal Republic of Germany gave me the honor of asking me to be here today among you, and to bring my modest support to Mr. LaRouche in the fight he leads against the Justice Department of his country, I did not, for a single instant, hesitate. I thought in effect that this was an opportunity to shed light on an attitude and a belief that is common to us.

Surely, it is not easy to find words that are appropriate to define what is fundamental that I have learned through the themes developed and courageously defended by Mr. La-Rouche and the Schiller Institute. But, to the extent that communication requires the use of words, I have hazarded into a tentative definition which has only two words: a *humanist radicalism*.

But what must we understand by that, and what bearing does the qualifier "humanist" bring?

I would freely say that this "radicalism," first of all, does not consist in being isolated to a certain category of ideas, but rather it implies a particular attitude, an approach. But before all reflection, this radicalism leads us to adopt a motto: *Omnibus dubitandum* ["Placing everything in doubt"]; for, we

Mr. Mpinga Kalongi *is a representative of the Embassy of Zaire in Bonn, West Germany.*

67

must, at any given moment, place everything in doubt, and in particular, all ideological concepts that are so well anchored in public opinion, and which—unfortunately—are on the road to being transformed into axioms, without appeal to all the caution of good sense that gives them meaning.

But this does not have to do with that doubt which is synonymous with that spiritual malaise that is unable to arrive at any decision or conviction. Rather, it has to do with that ever-present possibility of subjecting to criticism established postulates, sanctified by common sense, logic, or what is called "the natural order of things"! This "radical" calling-into-question is possible only to the extent that we do not take for self-evident received ideas or concepts.

Mr. LaRouche and the Schiller Institute, it is true, demand we enlarge our field of vision while compelling us to limit the influences, conscious or unconscious, that work on our own reflections. It is in this fashion that this doubt would enable us to both unveil and discover. We must, as in the fairy tales, perceive that the emperor is naked and that his "splendid" robes have been woven only by illusion.

In effect, this doubt planted within us aims to pose again the challenge. From this standpoint, it cannot be considered as a negative process. Nothing is easier than to contrast a reality to its opposite. This *"radical"* doubt remains, it is true, dialectical, for it upholds itself through the harmonious development of contraries, that is, it aims toward a synthesis that both denies and confirms. It is, in my mind, the instrument for the liberation of thought and systems still enthralled by their idols. In effect, it permits an enlargement of the field of consciousness, a more imaginative vision, and the creator for our potentials for choice.

Far from working in a vacuum, this *"radical reflection"* springs from the *"essential,"* that is *man,* wherein he is the point of departure and the point of arrival. But to say that man is the essential should not lead to taking this affirmation in its positivist or descriptive meaning. We must not consider man as a *thing,* but as a *process of becoming*: For he has the potential to achieve, to affirm, to derive benefit from an equilibrium that is greater, a love that is more intense, a universal consciousness more open.

We intend also to speak of man as a being whom one can corrupt, where the power he has to act can be perverted into an appetite for the power to do evil or to act against another, just as his love of life can become changed into a perverse thirst for destruction.

This humanist radicalism is a *calling into question,* which both perception of the profound forces of human nature and the concern for the growth and flourishing of man guide. Contrary to contemporary positivist thinking, this radicalism is not *"objective,"* if we hold that objectivity consists of latching onto theories, without making an appeal to an impassioned ideal that fortifies thought and incites it to go forward. We speak of objectivity and we mean by that, that each stage of reflection is founded on proofs, carefully subjected to critique, and that, above all, that we subject to examination the common sense premises. Once more, what this means, is that this radical humanism puts to the question every idea, every institution, every political, cultural, economic, scientific system, etc., in order to find out whether they serve or obstruct this attitude where man seeks to live in plenitude and joy. It would take very long for me to submit to examination all the examples of this kind of common sense premise that this radical humanism calls into question. But, nonetheless, I would like to bring up and underscore a situation which, for my part, assumes a capital importance, be it only by reason of its ever-burning presence. It has to do with North-South relations.

## North and South

In effect, never, it is true, has the South, or more exactly the Third World, been so intensely evident in our daily lives. It is, of course, in the gasoline that makes the engines of your cars run and it is also in the body of the cars. It is in the furniture that decorates your living rooms, and in the store window displays of early season vegetables before they enrich your daily menus. It is also, and frequently, the recourse of your publicity campaign and vacation plans.

It is in the copper in your household appliances and its silhouette is unmistakably outlined behind your growth and inflation curves. Never has your economic prosperity and equilibrium of your balance of trade been so dependent on it.

But since forever, in the different discourses from economists in the countries of the North, the Third World, or precisely the South, it seems to me, has always been presented as a barbarian to be civilized. And aside from several good adventurers or generous spirits, few among you are those who have strived to really study the fate of the South and to become interested in the very nature of North-South relations. For many, it is true, this question comes only to mind on occasion as fearful images projected from the TV and which cuts short their appetite at table. For others the Third World is the hand outstretched for their "generosity": This "generosity" which permits them to advantageously furnish international conversations, and from time to time to fill in a few lines on an electoral program. That also permits them, it is true, to not speak of any other matter, and notably of the "aid" which the countries of the South bring to the development and the growth of the countries of the North, since the structural mechanisms that preside over the relations of the two hemispheres are so integrated into the order that they have become practically invisible. We must denounce this, shed a bit of light on it, and force it onto the agenda of international conferences.

The classical theory of foreign exchange supposes, in effect, that the raw materials requirements of the countries of the North, and the scarcity of capital in the countries of the South, bring a significant flow of capital into the latter. This flow would normally compensate for the negative consequences of foreign exchange rates on the development of the countries of the South. But it is nothing of the sort.

Private capital has dried up, and, as a result, the economic dependence of the countries of the South is increasing at the very moment that public capital is pursuing political rather than economic objectives.

Public aid ought normally to take the place of private capital and to facilitate thereby world development as a result of capital whether under grant, or under lending at long-term, but generous, conditions. Moreover, bilateral aid that a rich country furnishes to a developing country should be succeeded by multilateral aid. In this form of aid, an international organization makes an aid grant thanks to the contributions of rich countries. Such a form of aid will have the advantage of better separating *aid* and the *political strategy* of rich countries.

To become industrialized, the countries of the South are obligated to become indebted. Bit by bit, the weight of their creditors in policy-making grows. Of course, the creditors are themselves more and more tied to their debtors. The interruption of payment from an important debtor could place the entire world economy in difficulty, for the global banking network has woven the threads into a mesh that is ever tighter and more constrictive.

## Preconditions for Development

The industrial development of certain countries of the Third World is in fact possible. But it is not therefore possible to have development without a profound transformation in the relations between the currently industrialized countries and the others. There are several means to envisage this change, but it presumes more than a simple reform. An overhaul of the world geopolitical equilibrium is necessary. This must emerge into a new type of "North-South" relations, or into an aggravation of conflicts and present contradictions.

One of the principal exigencies of the countries of the South has always been the improvement of the market structure for raw materials. They hope for a substantial increase in their export receipts and in the revenues of their producers, which, consequently could have positive effects on the revenues of the Third World. By definition, this will stabilize and eventually upgrade the revenues. Hence, it will help the countries of the Third World to, themselves, resolve their own problems in the balance of payments, of debt, and of exchange rates, together with helping them to successfully conduct their fight against famine, illness, and underdevelopment.

The problem of reorganizing the international economy, like that of the domination weighing on the Third World, has brought us to a certain fact: *We must call into question the growth of the countries of the North*; which does not in any way mean to suppress it, but to reorient it. Are we able to do it? The entire question lies in the good will of each person.

However, if your societies in the Northern Hemisphere continue to become enmired in imbalances and aberrations in a growth whose failings are more and more patent—it is not impossible that one day, the countries of the Third World, who want to break the vicious circle of misery and famine, would join to put the growth of the rich countries on the back burner and force the latter to make amends and reorient it in their favor. Otherwise, their development will forever find itself compromised or risk being nothing more than solitary and antagonistic in place of becoming interdependent. Without that, the world is making ready for new and grave confrontations.

Similarly, if the leaders of the developed countries do not perceive the exigencies of the present situation, they will without doubt witness phenomena in the internal affairs of their own countries, of social and political deteriorations capable of completely blocking the already complex mechanisms of their industrial societies.

Your responsibility is immense. You must help the weak economies and civilizations become interdependent: You must bear witness that economic efficacy can be joined to, and reinforce, respect for others, and cause the growth of their dignity.

That is why I call upon you to discover together what we must do to place the power of man at the service of man, of dignity, and of the joy of each among the peoples of the world.

It is in order to be living in the future that we have convened here. Let us unite in joy to uphold and celebrate this matter of conscience for which Mr. LaRouche has put himself forward as the standard-bearer.

JORGE PANAY

# Panama's Fight for Sovereignty

Mr. President, Mr. LaRouche, friends from the United States, friends from around the world. I want to convey to you the greetings of our people, from a people today fighting a hard battle against foreign forces trying to dominate them and continue the colonialist process. It is important for us to be here today, because life's experiences have taught us that only in the field of international solidarity of sharing ideas and joining ideas from other peoples could we find solutions for our problems, which, in most cases, are equal to those of other peoples of the world. This morning, when we heard what Mr. Lyndon LaRouche had to say, we felt really happy to find intelligent people in the North, leaders who share and understand our problems. This is important for the younger generation of Panamanians.

We believe that united we could win, win for a better and more balanced world. Panama is a small country, with only 2 million inhabitants, who are linked in a crucial way to the world economy. The Panama Canal was built across our land. Until 1977, it was controlled by the U.S. government. Since the treaty [in that year], the Canal has been jointly administered, and in the year 2000, this Canal should—despite everything—become totally the sovereign property of the Republic of Panama. A trans-isthmus oil pipeline also cuts across our country, carrying U.S. petroleum. Our country contains

Jorge Panay *is a Panamanian economist.*

73

a trade center linked to the world, where products mostly from Japan, the United States, and Europe are sold. Our country is also an international banking and financial center, linked to the biggest world interests. Our country has a barely developed industry and an agriculture undergoing ever greater problems.

It seems that with all these links, all the presence of interests and capital from the dominant economic power centers in our country, our population's unemployment levels, poverty levels, and housing problems have been growing.

The economic growth problem has been aggravated by an economic war unleashed against our country by . . . the United States. The foreign debt impedes our development process, by looting huge amounts of resources needed by our people. To solve all these problems, we believe we should join forces to change the international economic system. We believe the solution to these problems is of a political—not a technical—character. The problem of lack of capital, the debt problem, clearly described during this entire seminar, should be resolved by political paths, and jointly, with the participation of each and all of the nations affected by it: real participation, sovereign and independent participation.

We suffer from unequal terms of exchange. . . . That should be solved politically, through agreements between countries and between their men. The lack of cooperation by the more developed countries toward the less developed has to be solved, because the problem belongs to both sides. If we do not develop, if our peoples' problems are not solved, the developed countries will not be able to solve their problems. And, every day, things will get worse, and we all will suffer.

We should also jointly solve the problems of political aggression. I have to take advantage of this tribune to reveal that our country and our government are under fire from a conspiracy run from Washington to destroy our government, destroy our people, destroy the conquests we arduously won in the anticolonial battle for sovereignty and national independence.

We have been the direct victim of conspiracy and blackmail from the World Bank and the [International] Monetary Fund.

Their policies, their conditions, are increasing poverty and unemployment among our women and our men. In Panama, 38 percent of the women have no jobs. Twenty-four percent of men under 25 years of age have no jobs. The boys who graduate from high school and from college cannot find jobs. And this is the direct result of the application of economic policies which are alien to our reality and alien to our problems. The impositions and blackmail by certain sectors of the Reagan administration against our government must stop. In that regard, in taking advantage of this tribunal, I want to ask the members of this conference to help us announce our situation from every podium, everywhere you go in your countries.

## Self-Determination for Panama

We have received solidarity in our struggle to recover our Canal and sovereignty. We need even more of that solidarity right now, so that the forces of regression do not take back the reins of our nation. We have been open in our international relations. We have respected all countries. We believe in self-determination.

We have been clear on the Central American problem. We want peace. We do not want military intervention from any place, from any other country, because we Central Americans want to find our own answers, our own solutions to our own problems. In this regard, Panama has made every effort to participate in the Contadora Group. We have made every effort to participate in and promote every Latin American peace initiative. We are a Third World country because we believe in the people who have problems like ours. We have sought to integrate our markets, integrate our economies. We believe that to be able to solve the problems inside our country, we have to seek international aid, in the form of technology and capital to develop the areas which we recovered through our sovereign struggle for the recovery of the Panama Canal.

We believe that when, in the year 2000, the administration of the Panama Canal returns completely to our country, there must be a design which tries to integrate Latin America, which tries to integrate the world, a design for the Canal's administration and operation to be shared with all countries in order to develop and raise the living standards of all humanity.

In this regard, we have to protest the unilateral violation by the U.S. government of trade agreements and of agreements on personnel administration and political accords agreed in the Panama Canal Treaties. We know that from now until 2000, we are going to run into some problems, that many forces which don't believe in change, many forces which don't believe in our peoples, are going to oppose our taking the Canal in our hands, oppose our developing ourselves and consolidating our political goals, our goals of democracy, our goals of national liberation, of respect for other peoples.

But we also know that, as we have shown here today, every day other men from other peoples, white and black, in every country on earth, every day more agree with our interests, agree with our political ideals of development and respect for all mankind.

# Update the Canal

We also clearly need help to study the alternatives for the Panama Canal. The current Panama Canal is losing its technological value hour-by-hour. It is going to be obsolete in the future. We Panamanians have to introduce technological changes in that means of communication so that it continues serving the development of humanity. There is currently a tripartite commission made up of the governments of the United States, Japan, and Panama seeking an alternative for the Panama Canal. In our view, more countries should participate in that search and for the solution to be decided in function of world development, raising mankind's living standards.

To conclude this speech, I would like to give you a message from our people to the organizers and those present: We, in Panama, trust in you; we, in Panama, trust that the world will change; we believe the change will be positive; we believe the world is smart enough and has enough creative capacity not to collapse and to reach the year 2000 with better living standards for each and all the inhabitants of the planet Earth. Thank you.

PATRICIO ESTÉVEZ

# The Economic Crisis in Mexico

Mssrs. LaRouche, Lyndon and Helga, friends gathered here from many countries:

I want to outline the new effort being made in Mexico to reverse the rapid economic and political collapse going on for the past five years. The decisions Mexico takes in these pre-election months must change the country's path or there could be an historic regression with grave consequences for the future. As all the Latin American countries, Mexico is assaulted by serious foreign pressures, by the enormous weight of its foreign and internal debts, and by a worsening of injustice, because of deviation from the goals and intents of the Mexican Constitution and Revolution.

During the past five years . . . the class which governs the country has seriously increased economic dependency by following all the dictates of the International Monetary Fund, which has caused brutal impoverishment and a permanent threat to the individual freedoms guaranteed by the Constitution. . . .

Public enterprises have been dismantled, thus reducing the material base of our sovereignty and fulfilling the fondest dreams of foreign interests. The entire burden of the crisis has been dropped on the shoulders of working people. The productive apparatus has been denationalized. Internal con-

Patricio Estévez *is a representative of the Authentic Party of the Mexican Revolution to the state legislature of Sonora, Mexico.*

sumption of the Mexicans and their hopes for a decent life have been reduced. . . .

The economy is being opened to a new neo-colonial phase, intimately linked to the in-bond workshops on the U.S. border and exports of consumer goods to the U.S. market. This has made it very hard for small and medium industry to survive. The value of the labor force has been cheapened and the neo-colonial model could become irreversible if . . . the about-face under way were to destroy the nation's productive possibilities.

Servicing the impossible debt has been accomplished at the expense of economic growth and popular well-being, and has caused the Mexican workers to lose more than half of their buying power during these five years.

Inside Mexico, a small privileged sector multiplies its acts of financial speculation through the stock market. It has blood-let so much through capital flight. . . . The chain-letter refinancing of debt payments, of illegally contracted interest, has been so infernal during these five years that the country has already become a net exporter of capital. . . .

## An Emergency Economic Program

Against this situation is a new phenomenon of unity of the republican and nationalist anticolonial, anti-imperialist forces of citizens. This is basically made up of the middle classes, ever-growing parts of the working class, and the deeply decapitalized agricultural sector. They are fighting to turn around the situation by means of today's crucial electoral period.

In broad outline, the program of this new movement primarily seeks to facilitate, as set forth in the Constitution and the laws, full respect for popular sovereignty as expressed by the vote. It proposes changes in election and political laws to make suffrage truly effective under these conditions. In the electoral area, it poses the urgent need of stopping—no matter how—the rapid impoverishment of the majority of the population. It looks to an emergency economic policy which puts an effective halt to inflation, immediately raising internal consumption standards, and promoting employment by redirecting credit to productive purposes and away from speculative activities.

It is urgent that the supply of basic goods, which day by day become less available in Mexico, be increased. Interest rates, already in the triple digits, must be decreased. Price increases for all consumption articles and agricultural and industrial inputs must be stopped. The slide of the peso must be stopped, in order to replace the now-hegemonic speculation with productive generation of socially useful wealth.

I agree with the speakers from other countries that international economic and financial relations must be changed, with debt service suspended in whole or in part, and debt payments adjusted to sovereign decisions on the economic processes to be taken to reconstruct the country. Public and productive companies must be saved. We must suspend debt service until equitable decisions are made to correct the size of the principal, to reduce interest rates, and to limit payments to a small percentage of our export income, after satisfying the needs for national development. New debts to pay old ones must be prohibited along with letters of intent. . . .

There must be a change in the foreign investment law, which for years has brought violations and has been one of the major sources of capital flight through conversion of pesos into dollars, as a kind of speculative savings which has caused immense economic bloodletting.

The economic reconstruction program should divert most of the money liberated by not paying debt service toward productive investment and development. Economic policy must be oriented in terms of industrial priorities of production of basic consumption and agricultural goods.

The creation of agricultural and irrigation infrastructure must be freed from present policies which have paralyzed it. What Mr. Frank Moss said about this a few hours ago was, for Mexico, a valuable way reformulating the international relations of our country with our neighbor to the north, the United States.

Patricio Estévez

# Farm Sector in Collapse

Mexican agriculture is extremely decapitalized and is in a crisis. It has been hit this year by high interest rates, high prices for inputs, delayed credits, moratorium interest, and finally, the freezing of so-called price floors. The result today is that the harvest is bringing in less than its production costs, while the burden of domestic debt on agriculture is getting much heavier.

The production cost per hectare of wheat rose from 1980 to 1986 by 2,600 percent in current pesos (from 7,540 to 190,000). During the same period, however, the guaranteed floor price only rose 1,600 percent. This decapitalization machine has harmed farmer living standards. What is worse, is that with each crop cycle the gap has increased between production costs and the prices paid for popular consumption products. Thus, the government has resorted to the suicidal policy of continuing importing basic foods, mostly from the United States. I'm talking about corn, beans, and other grains.

. . .In 1980, it took 300 tons of wheat to buy a harvester; today it takes more than 1,000. Therefore, in the irrigated areas, most of the equipment has not been replaced and is far past its useful life. . . .

Mexico should suspend debt service to give urgent priority to dams and irrigation systems which have been paralyzed for many years. New areas can be irrigated by channeling the 90 percent of the water in Mexican rivers which now flows, wasted, into the sea. There is no way to get these partially built projects going again so long as the government keeps implementing IMF austerity policies.

Fifty-six percent of this year's Mexican federal budget is condemned to be handed over for servicing the foreign debt; but only 3 percent goes to agriculture. The urgent waterworks to be reactivated include the Pacific coast plan and the Gulf [of Mexico] coastal plan. The NAWAPA, the irrigation plan between Canada, Mexico, and the United States must be implemented. If it is done on just and equitable terms, it could become one of the main pillars of a genuine alliance between Mexico and the United States. If we can bring that alliance forward, both in Mexico and in the United States, we will be giving historical continuity to the alliance which a century ago linked President Benito Juárez with President Abraham Lincoln, when both republics were threatened by the empires.

Thank you.

FEDERICO SOSA SOLÍS

# The Need for a Debtors Alliance

Greetings.

I am a lawyer in the state of Yucatán, the Republic of Mexico, where there are Maya vestiges which give us fraternal links with the tourists who visit us over the years. I am extremely happy to come before you to cooperate with your struggle, by means of the Schiller Institute, which organized this conference.

The New World Economic Order sought in this conference now means not only a proposal of the debtor countries, but of humanity itself, which is today seriously threatened by hunger and nuclear war. The historic image of the enchanting village of Bretton Woods, where economics experts met in 1944 to plan aid to the peoples devastated by the World War, is present in this forum. It is present here in Boston, where intelligent men and women are meeting to lay the basis for a New World Economic Order. It is relevant that this takes place in one of the main cities of New England, the historic and cultural root of the great people of the United States of North America, because the first colonizers with the banner of freedom arrived here.

The foreign debt problem of the Ibero-American countries can only have a solution in favor of the debtor countries by means of an association of them as an economic and financial bloc in the face of the powerful creditor banks. Those countries

Federico Sosa Solís *is a lawyer in Yucatán, Mexico.*

83

which have tried to find a solution by themselves, without unity with the others, have had only negative results. Their economies have been injured, because the renegotiations imposed on them extremely high interest rates and interest upon back interest.

They were forced to sign letters of intent, accepting in almost all of them debts equivalent to double the real amounts they owed. To give an idea of these injustices, I am going to give you numbers on the increase of the combined debt of the Ibero-American countries, numbers provided by the Schiller Institute itself. In 1977, the foreign debt of these nations added up to only $123 billion. That figure itself was already triple their 1973 debt, which was only $40 billion. But, due to the unjust renegotiations of the last few years, the total debt is now far more than $360 billion.

## The Role of a Common Market

The debtor nations, to deal with the difficult situation afflicting them, have no choice but to integrate themselves in a new economic order based on Latin American unity and the creation of a common market which preserves the real value of the products competing in that market. The common market would rapidly become quite large, thanks to the diversity of products in it.

Unity with whom? in what?

Unity is not simply an ideal, but a well-defined objective since the beginning of the last century, when our peoples achieved their independence. In those days, their leaders thought of the whole, of achieving a greater fatherland, based on the historical, linguistic, and cultural roots which unified them, and the geographic continuity of America's territory. Future borders would be mere territorial demarcations, which would serve only to identify the families of one great Ibero-American people.

We must be very frank. Interest on Ibero-America's debt is unpayable, is uncollectable, because the economic and financial resources of our countries will not permit these enormous debts to be paid—even with the greatest sacrifices and privations. The only thing the international banks could achieve would be to socially destabilize the debtor countries, with horrendous effects not only on us, but also on the creditor countries. It would kill the goose that lays the golden eggs, but at the same time, the potential social explosions would have direct repercussions on the life and stability of the rich countries, because they could not survive without the existence of the underdeveloped countries, which are their sources and which nourish their great economic structures.

If we did not pay interest on our debt, it would cause the crumbling, the collapse, of the world's financial system. It would lead 80 percent of the banks in the rich countries to an economic collapse—and the creditor banks know this. . . .

A calm analysis of the problem leads to proposing one of the measures necessary for the economic survival of the debtors, which consists of negotiating new loans with fresh money, but without interest payments. The fresh money loans would permit the debtors normal growth. They would repay these loans with percentages of their exports. I repeat, no interest of any kind, except in the long term.

This option is the only one which the debtor countries could present and is the only one which the international banks would have to accept in the face of world realities. Therefore, we must be firm, and present our points of view forcefully and seriously. I think there is no other way the banks could be repaid their previous loans and the new ones they would give us. It would permit a genuine conciliation between the rich and poor countries with the establishment of a new economic order for Ibero-America and for the Third World. This would facilitate a peaceful coexistence for a long time; since, if it were not done, democratic systems would disappear from the debtor countries, with what follows in the social, political, economic, and financial sectors.

## A Limit to Austerity

The International Monetary Fund's recommendation to apply austerity is incorrect from the social and realistic point of view, because there is a limit. That limit is the peace and health of the peoples, as has been proven in those countries which have imposed it. . . . [IMF policies] neither reduce nor stabilize foreign debts. On the contrary, those countries who follow them have greater need for foreign capital. . . .

There is some truth in the statement that debt and austerity are measures to pressure debtors. This kind of behavior is neither healthy nor moral, since it brings with it things which harm the political and social organization of the peoples of the Third World.

In order for this great people of the United States to know close-up the way Ibero-America became indebted, I will summarize: The burden of direct capital and capitalized interests added are due to higher interest rates than those of 1977; the effect of the deterioration of terms of trade; capital flight; and interest paid on new debts resulting from the previous two problems.

The creditor banks' tactic has been to get the debtors to accept renegotiations under conditions of extremely high interest rates, double or triple those before. The interest rate increase which began in 1978 and continued in 1979 was fatal for the economies of the debtor nations, even for those like Brazil which have trade surpluses. Countries could not pay these rates, resulting in the enormous total of interest bringing an accumulation of debt. . . .

By 1983, Ibero-America's debt became unpayable, leading to national solutions, without any of them being a solution to the problem, because Ibero-America and the countries of the Third World would not present a solid bloc with clear and firm decisions. There are charts and graphs produced by expert economists showing that if there had been no price changes on exports and imports, the debt of the underdeveloped countries would be a large percentage less today. They calculate that deterioration of terms of trade caused losses of more than $95 billion. If that amount had been used to pay the foreign debt, it would have greatly diminished the economic problem studied in this conference.

As a logical result, living standards in the Latin American countries have fallen visibly, given the austerity measures imposed by the International Monetary Fund. This has also affected real wages, with lower buying power resulting in lower production of durable consumption goods.

Since the gross national product of the conjunction of the Latin American countries has systematically diminished during the last few years, and this makes it impossible for any Third World country to develop, the Ibero-American countries face an uncertain future.

The moment has arrived to make firm and decisive decisions. The international banks [must realize] that interest on the foreign debt is already unpayable, and that capital will only be paid through small percentages of exports. This is the posture of the New Economic Order we, at this historic conference, are designing and beginning. Thank you.

ZORAIDA ELSEVIF

# The Results of the International Monetary Fund's Neo-Malthusianism

Ladies and Gentlemen, man is the principal resource an economy counts on for its development, because he is the only active factor in the productive process. He is the origin and, at the same time, the end of all economic systems. . . .

. . . That is what brought the marked growth of our country's foreign debt, which rose last year to more than $4 billion, almost five times its level at the end of the 1970s.

That indebtedness, however, did not go to strengthening the productive sector of our country, nor to favoring the population's needs. That indebtedness, gentlemen, by agreement between our dominant groups and our creditors in the international financial system, financed—in the majority of cases—the high prices of imported components of the also imported technologies and the imported products which come to our countries.

And who does this really favor? Finance capital in the developed countries through interest payments, consulting fees, capital purchases, etc. And once again, to service the debt itself. . . .

Today, the International Monetary Fund, through its so-called "adjustment policies," intends to contract our already strangled economy. The proposals to devalue our money and

Zoraida Elsevif *is an economist from the Dominican Republic.*

## Zoraida Elsevif

create a free money market are mechanisms to reduce real income, to ruin the buying power of the Dominican peso and reduce internal demand, and thus, the Dominican's quality of life.

It is true that these International Monetary Fund recommendations would tend to help reduce today's trade balance deficit. But, it would do so at an excessively high cost for us, because it would bring hunger conditions to large parts of the population which are already earning the minimum and live merely from underemployment. But, the [IMF]'s intentions do not stop there. They carry out their neo-Malthusian policies in the long term, because the impact of the "adjustment" measures on our economy's incipient productive activity will result in either greater dependency or greater marginality of man. . . .

# II

## The Dignity of Man in a New World Economic Order

# The Dignity of Man in a New World Economic Order

I would like to speak on the question of the dignity of man in the New World Economic Order. The question whether humanity will survive this crisis or not, depends upon our image of man, and on the ethical standard which flows from that image of man. The prevailing ethical standard, in turn, determines political and economic developments in any given epoch.

In his new encyclical, the Pope presented a thesis which on first inspection might seem quite amazing, namely, that even though the pathway to eliminating underdevelopment is well known—not the least through the encyclical *Populorum Progressio*—yet, the trend in today's world is not toward development, but toward death: "In today's world, including the world of economics, the prevailing picture is one destined to lead us more quickly *toward death* rather than one of concern for *true development* which would lead all toward a 'more human' life, as envisaged by the encyclical *Populorum Progressio.*"

This, unfortunately, is the face of reality. Indeed, when we look at the current course of humanity in all its political and

Helga Zepp-LaRouche *is chairman of the Executive Board of the Schiller Institute. This was the keynote address of the second day of the conference, March 27, in Cologne, West Germany.*

economic diversity, we can only say that what we human beings are doing with the world handed down to us, is nothing less than suicidal. Each new day in our current affairs seems to bring us one more step toward destruction, toward death. Perhaps for the first time in the history of the human species, we face the prospect of the demise of all human civilization, with the human species engaged in activities which will lead to its own destruction. And because of the international interdependence between North and South, East and West, it will not be only *one* civilization or culture which is destroyed. The crisis in our civilization has far greater ramifications than, for example, the Roman Empire, which collapsed for very similar reasons, namely, the turpitude of the masses and the decadence of the ruling elite. Indeed, the Roman elite was so decadent that it was unable to even recognize the threats encroaching on its system from without and from within. Their preoccupations lay entirely elsewhere, in revelries, burning Christians, orgies—entertainments which titilate today's jet-set as well. . . .

## What Are the Structures of Sin?

When the Pope speaks of the "structures of sin," this is a relatively mild term. It is these "structures of evil," which will be responsible for the destruction of humankind. The basic problem is the same as with all previous periods of decline: The ruling elites have an image of man which places no value whatsoever on the human being. Children dying in Africa—who cares about that? It's scarcely acknowledged to exist. Tens of thousands, hundreds of thousands of human beings can be dying of hunger—and once in a while people see these awful pictures on the television, and then switch the channel. What kind of people are these, who can look at these tragedies without something rearing up inside them, and each saying to him or herself: "I can not morally reconcile myself with this—I can not tolerate that so little value is placed on a human life"?

If a citizen of a developing country is so unfortunate as to get run over by an automobile, it often takes hours for help to arrive—simply because there's no money for ambulances. For those who are responsibile for the misery afflicting large parts of the globe, such human suffering only brings up indifference—an indifference which scarcely conceals their boundless selfishness. Their own possessions, their career, their family are the first order of priority; nothing else really interests them.

To put it another way: These injustices in the world are not natural catastrophes, but are the result of concrete acts, and failures to act, by individual human beings. And no man can get around that fact. They are the result of specific, concrete policies. Imperialism's image of man—in both its Western and Eastern varieties—is predicated on precisely these egotistical structures. At the center of policymaking, there is no place for man, but only for self-serving aims. In many cases, man is considered more useless than livestock, since livestock can be slaughtered or milked or can lay eggs, whereas human beings are merely useless eaters—as has been made explicit in the euthanasia movement which, unfortunately, is now re-emerging.

How can we look on idly, when in the allegedly prosperous United States alone, 400,000 children are homeless; when in Brazil, 40 million children live in the streets; when the financial system in Italy alone is propped up by $80 billion in drug money each year; when the drug mafia has become an essential mainstay of the world economy? The result is, that children's minds are destroyed. Already in the schoolyards of American cities, six-year-olds are hired by twelve-year-olds to stand on the corner to work as the drug pusher's extended arm, because the latter doesn't want to attract too much attention. The process of degeneration in the United States is so brutal, that many parents no longer send their children to school—out of fear that they might get ensnared in this infernal machine. What is it, inside us, and in the political system, which tolerates all this? It isn't as if no one knows about it. The facts are well known.

Clearly something is wrong with us as a society in general. Another example is the video market. It has been reported to me, that nowadays people go to the video store with a plastic bag—because at least they want to conceal what they're carrying—and they check out twenty videos for the weekend—and what garbage! Ugliness! Every once in a while, I admit, I've looked at American television, and it's unbearable! There are shows which are so ugly and full of brutality, and which convey such a base and ugly image of mankind, that any healthy human being who wants to stay human, simply can't look at them without destroying his own personality. Recently, the press was filled with reports that the organized practice of satanism and occultism is widespread in the secondary schools in West Germany, England, and elsewhere. That, in my view, proves that we are not merely speaking about the structures of sin, but about the personification of evil itself.

And at this point, I would like to present my thesis: that evil, or satanism, has played a much greater role in human history than the official historians usually admit. It has certainly been said often enough, that Hitler, Stalin, and Pol Pot were such incarnations of evil. But I am increasingly drawn to the conclusion, that satanism has played a continuous role as a history-shaping force. This continuity is visible to anyone who wants to see.

# Devil Cult and Occultism

The ideology being promoted for the new world dictatorship is the so-called "New Age"—in plain English, a philosophy which asserts that we no longer need rationality, but only to "feel." The age of Christianity, the Age of Pisces, has come to an end, and must yield to the Age of Aquarius. No longer is there progress in the world, but instead a return to the old cyclical cosmology of the pre-Christian mythologies. This "New Age" is not all that new: Nietzsche, Dostoevsky, and Aleister Crowley are counted among its spiritual forefathers, and their teachings led directly to Bolshevism and fascism. Today, all

Helga Zepp-LaRouche

this is sold in a new package, and even though it's called "New Age," there's really nothing very new about it.

It can be demonstrated, that whenever occultism assumes a dominating role in a society, that society collapses unless it promptly changes course. The fourteenth century is one example of this. But the moral crisis we are undergoing today is deeper than it was then.

Let us consider the so-called power structures—the structures of sin—and let us delve more deeply into what is lurking underneath, and into the ethical principles on which they are founded. The mainstays of these power structures strive to uphold those structures at any cost, for the sake of their own egotistical motives. Now, the question whether humanity further develops, or is destroyed, depends upon the emotional attitude of individual human beings, and on the ethical principles accepted by society at large. So, I would like to give a name to the ethics of the phenomena I have just described: the ethics of hate. The people who wish to maintain these structures are entrusted with the task of producing wickedness within others. And unfortunately, the devil is very clever; as in Goethe's *Faust,* he comes with all sorts of enticements. He doesn't picture hell as it really looks, after the fact: the ovens where you are roasted, the pitchforks which tweak and torment you. Rather, he seduces his victims with the great variety of "freedoms" they can supposedly enjoy, in the realm of sex, in "the easy life." What the Bible says is unfortunately all too true: "The name of Satan is temptation." It begins with the littlest things, but once you have set off along this path, you are led ever more rapidly downward. There exists no such thing as equilibrium, a steady state—not in physics, nor in history, nor in your personal life. Either the person strives to perfect himself, or he must yield to these temptations.

The people who rule today's power structures were acting quite deliberately, when they sent cultural pessimism into the world. One exemplar is Erhard Eppler, who asserts that we have reached the limits to growth; or the Club of Rome, which has robbed young people of any hope that they might strive for something higher and better than today. It is relatively easy to rob people of their hope and to arouse the bestial side of their personalities; and unfortunately, the past twenty years' slide toward decadence in our society has completely borne this out.

These powerful individuals act according to a brutal ethic. As long as you don't disturb them, they won't hurt you. But if you begin to say, "No, I will fight for other ideals—for the industrial development of Panama," for example, then a General Noriega has to fear for his life. Recently, a U.S. government official bluntly stated that, "If Noreiga does not go voluntarily, he'll soon be leaving Panama in a coffin." The ethic of these people, when their own self-love is challenged, is overt hatred, and stark brutality. Their highest precept is, "Don't get caught." And it is also in their ethic, that it's a crime to permit outsiders to peek into this underworld society.

I would now like to answer the following question: Do these people possess anything, really? Should we envy them because they are rich? I adopt the standpoint of Schiller, who considered these people to be the poorest of the poor. Elizabeth says to Don Carlos in Schiller's so-named drama: "How poor, how beggar-poor hast thou become, ever since thou lovest no one save thyself." And Schiller says in general, that whenever I hate, I am taking something away from myself. Hatred of one's fellow man is nothing but protracted suicide, and egotism is the most impoverished state of any creature. So these parasites, these leeches, who could not care less if their entire plundered system is destroyed, are actually not so especially well off. Despite this, of course, we have to realize that a continuation of a policy which leads to an ethic of hate, and to the further violations of the laws of Creation, will give truth to the encyclical's statement, that at the point when nature shall no longer recognize man as its ruler, the human species shall become extinct.

I have not gone through all these things, just so I can leave it at that. Our only purpose in recognizing these dangers, is that we may become increasingly determined to do everything to avert them. Personally, I do not believe that the plan of Creation destines man, the crown of this Creation, to become a beast. Rather, I believe that man has been given free will in order that he emerge victorious over false structures. This is why we must fight, with all the strength we can muster, to reintroduce an image of man which is based on the principles of Christian humanism. And if humanity is to survive, this image of man must be in harmony with the laws of the universe.

## The Laws of Nature Are Intelligible

In Christian humanist philosophy, these universal laws signify nothing else, than that the inviolable dignity of man is grounded in natural law; i.e., by virtue of the order of Creation itself, every individual born into this world has inalienable rights, which must be guaranteed, and whose basis lies in the divine nature of each human individual. The order of Creation, the physical laws of the universe, are negentropic (modern natural science has yielded sufficient proof of this); i.e., the physical universe is a negentropic process, in which successively higher manifolds develop, and in which evolution occurs without the intervention of human reason, but in which man, since his first emergence, represents the most developed creature in that universe. The human mind (as microcosm) operates according to the same lawfulness as the entire physical universe; and precisely because of this equivalence of human thought with physical laws, it is possible for human beings to progress forever onward toward truth.

Every human being who wants to contribute something to humanity's further development, is actually forced to replicate within his own mind, in a condensed form, the development of the entire universe. And so, Nicolaus of Cusa rightly said that since the human soul is the location where all new sciences are invented, the individual, if he wants to create anything fundamentally new, must have mastered all the essential questions of his era. That's what Nicolaus of Cusa already said back in the fifteenth century—so he was a very modern thinker.

Thus, man is capable of progressing indefinitely. He can progressively extend his mental conceptual powers into infinity. He can know more and more without end, because he is the living image of God's unboundedness. That is what Christian humanist philosophy tells us. No other creature besides man can improve himself through his own efforts. As nice as animals are, they are not able to decide one day, "Now I want to become a better dog," or, "I want to be a nicer cat or a thinner mouse." Only man can decide to improve and perfect himself by virtue of his own spiritual nature. The beasts, and everything else in the world, are what they are by dint of necessity, and only man's spiritual nature encompasses certain principles—freedom of intellect and will—by means of which he can become better, and thus more like God. When the mind comes to know itself as the living image of God, it also receives from God the power to become even more like His image, and to reach ever greater unity with those objects of truth which are characteristic of Him, says Nicolaus of Cusa. "It is up to me, that I become ever more receptive of Thee." And Nicolaus says that this receptivity comes from likeness, and that conversely, that unreceptivity comes from unlikeness. That is, man desires to become more like that which he loves, and he has no interest in that which he does not love.

Therefore, love is the power which expands the soul's power of comprehension for the domiciliation [*einwohnung*] of God. The more love is present, the more this likeness grows. Nicolaus describes this human capacity to participate in God, *capax Dei*, as follows: "Man is the image of the absolute art of the Creator, in that he can bring forth new things." The mind, in its knowledge, is turned toward the manifold content of the universe, and thus, as *imago Dei*, [in the image of God] toward the entirety of Creation—not only toward the Earth, but toward the entire physical universe. Therefore Lyndon LaRouche is correct, when he says that man's natural mission is to replenish the universe, and to make it into his garden. Our concern, of course, is that ever greater knowledge make man capable of ever greater knowledge; and the more man already knows and has elaborated, the more joy he experiences in knowing still more.

Helga Zepp-LaRouche

# Perfection Is Only Possible Through Love

Love is the force which makes this process of perfection possible—love in the sense of *agapē*, the love of God, of humanity, and of one's neighbor. Augustine said, *"Non intratur in veritatem nisi per caritatem,"* that is, "Man knows nothing truly without love," and this is absolutely true. Love always signifies a gaining of that which is loved, whereas hate always means losing the object which is hated. Love is the capacity to bring out the good within people. This is really the key to the entire problem. The Pope, in all his travels, has radiated *agapē* so strongly that wherever he has gone—the United States, Germany, Colombia, Argentina—he has always known how to tell these people precisely what means the most to them, thereby making them into better human beings, because they have been treated with love. Any human being who is not yet entirely driven by the devil, and who has not completely sold his soul, craves for this kind of love. And whenever these thirsting souls are touched by love, they become transformed.

The task of Christian ethics is to make man into the originator of Good. And anyone who looks at our great geniuses— Schiller, Beethoven, and others—or still better, is fortunate enough to observe such a genius in action—knows that these people always bring out the best in others, since they strengthen their desire to investigate, and so give them courage, and also strengthen their desire to become better people; so that the beneficial effect of these people basically lies in their ability to propagate Good all around them.

We are therefore entirely justified in saying, that Christianity, whose highest commandment is the commandment to love, actually leads to this ethic of love. The various biblical scriptures very clearly demonstrate this. John the Baptist was wont to say, "It is my command, that ye love one another." Paul said that the entirety of the law is fulfilled in one pronouncement: Love thy neighbor as thyself. Many other passages document how Jesus Christ called upon his apostles to feel love, and to go out in love. The scriptures also say that

101

DEVELOPMENT IS THE NAME FOR PEACE

love is expressed in practice in the keeping of the commandments, and is therefore not a vague generality. In the Pope's disputation with Max Scheler, a twentieth century theologian who gave primary importance to the imitation of Christ, the Pope countered that the commandment to love, and to imitate Christ, have two entirely different ethical contents with respect to Christian revelation; and that if one must assign priorities, one would have to rank imitation lower than the commandment to love, rather than the other way around. Love is, without any reservation, the greatest and highest of all commands. This means that Christianity, which is the basis of our Western Christian civilization and has shaped our history, sets up for us an ideal of ethical perfection, which can actually be visible to each and every one of us. And this idea has always characterized not only Christianity in general, but also the ethical and cultural high points of our own lives as well.

# German Classicism as the Expression of the Humanist Image of Man

The era of German classicism is indisputably the high point of German history, and it was based on the same worldview that inspired the Italian Renaissance and the periods of greatest flourishing in France. I would like to read to you a few words which Schiller wrote on this subject, in his *Philosophy of Physiology*:

"This much, I think, has been solidly demonstrated: The universe is the work of an infinite Understanding, and was drafted according to a magnificent plan. And just as it sped from draft to reality, through the omnipotent influence of divine power; and just as all forces act, and act amongst each other, like strings of a great instrument sounding together in a thousand-voiced melody: so man's spirit, ennobled by the forces of Godhead, shall discern cause and purpose within individual effects; and from the connection between causes and purposes, he shall discover the great plan of the universe; and from that plan, he shall know the Creator, and love Him,

102

glorify Him—or, which sounds more sublime to our ears: Man exists in order that he strive to attain the greatness of his Creator, and see the universe with the very same eyes as his Creator. Man was created in God's image; man's ideal, it is true, is infinite; but the Spirit itself is eternal. Eternity is the measure of infinity—i.e., he will continue to grow forever, but will never attain it."

Thus, Schiller supplied the answer to nothing less than the fundamental meaning of life. And this meaning is precisely what has become so beclouded today. Man fulfills his life's meaning, when, as Schiller says, he has taken Godhead into his own will. Schiller continues: "A wise man of this century said that a soul which is enlightened to the degree that it can keep the plan of divine providence completely within view, is the happiest of souls. An eternal, great, and beautiful law has tied perfection to pleasure, and displeasure to imperfection. Whatever brings the person closer to perfection—whether this be mediated or unmediated—will delight him. Whatever distances him from it, will give him pain. He will avoid what gives him pain, and will strive after what delights him. He will seek perfection, because imperfection causes pain; he will seek it, because it itself delights him. . . . And thus, it is the same as if I said: Man exists in order that he be happy, or: He exists in order that he be perfect. He is only perfect when he is happy; and he is only happy when he is perfect.

"But yet another equally beautiful, wise law—a corollary of the first one—has tied the perfection of the universe to the happiness of the individual, thus tying man to man, and even man to beast, with the bands of universal love. Thus, love— the most beautiful, most noble impulse in the human soul, the great chain of sentient nature—is nothing else than the exchanging of my own self with the essence of my fellow man. . . .

"And what is the purpose of this universal love? Wherefore all these pleasures afforded by universal love? Solely to serve the ultimate, fundamental purpose of fostering the perfection of one's fellow man. And that perfection consists in contemplation, investigation, and marvelling over nature's great plan. Indeed, all sensual pleasures (which we will speak of in due course), through their many involutions and seeming contradictions, nevertheless ultimately gravitate toward this same end. The truth will ever remain unchanging and self-same: Man has been created in order to contemplate, investigate, and marvel over nature's great plan."

And in his *Theosophy of Julius* Schiller writes: "Harmony, truth, order, beauty, excellence give me pleasure, because they put me into the active state of their inventor, their owner—because they betray to me the presence of a being whose sentiments are dictated by reason, and allow me to have a presentiment of my own affinity with that being. . . .

"Thus, love—the most beautiful phenomenon of all that is created with a soul, the all-powerful magnet of the intelligent universe, the fount of devotion and most sublime virtue—love is merely the reflection of this unique power, an attraction toward all that is excellent, based on a momentary exchanging of personalities, a trading of one's being for that of another.

"Whenever I hate, I take something away from myself; but whenever I love, my wealth is increased by that which I love. Forgiveness is the recovery of a possession which has been sold—and hatred of mankind is protracted suicide."

It is my view that the beauty of Schiller's image of man is unsurpassed, since he basically says that every human individual, on the basis of his inborn abilities, is gifted with genius. If the person works hard—and toward the end of his life, Schiller said that genius is diligence—and if he not only develops all his mental capacities, but also works to improve his feelings, his emotions, he can then become a beautiful soul. By "beautiful soul," Schiller means a person for whom there is no longer any contradiction between the rationality of reason, and the dictates of the impulses. The engineer who drafts a construction plan for a beautiful bridge, and then goes home and beats his wife, evidences a certain contradiction between rationality and emotionality. Only when the person can give free reign to his feelings without any reservation, precisely because the latter are on the same plane as his reason, does that person approach Schiller's ideal. Thus, Schiller's beautiful soul is a person who has allied his reason to the corresponding emotions, i.e., he is a genius, because only a genius possesses the freedom to extend law in an orderly way. And every person can attain to this.

Wilhelm von Humboldt, Schiller's very close collaborator and friend, poured Schiller's entire philosophy directly into his own famous educational system, which was established in order to foster all of the human being's creative abilities. According to Humboldt's educational ideal, the primary goal should not be to educate the individual into an expert in a particular field, but rather first to shape his character. That done, he can acquire all his other capacities in short order.

Provided we think about humanity in this way, all problems of policy, economics, and justice are really quite easy to solve. To propagate justice in the world, by seeing to it that every person unfurls this potential, this divine potential, this tremendous beauty, which he can then contribute to humanity— that is mankind's greatest mission. Mankind must establish a situation in which justice prevails, at least in approximation. This has very practical consequences: Only once man acts from this standpoint, will it ever be possible for him to find practical solutions. If, on the other hand, he starts out by bickering over the price, then it obviously won't work.

# The Strategic Implications of Economic Policy

I had the opportunity recently, in Paris and Rome, to present an analysis of Soviet warfare in the midst of peace time, from my standpoint as a years-long intelligence observer, based on the official strategic military doctrine of the Soviet empire. I find it therefore especially important, here in the United States, to address the currently acute, crucial problem of the credibility of Soviet declarations, i.e., words devoid of any evidentiary force up to now, taking into account facts which are proven and other matters about which we have secure knowledge. I believe this is particularly important, since Mr. Gorbachov caused the timely publication of his book, *Perestroika*, and on *glasnost* in summer 1987 in millions of copies, directed specifically at the American public. And also because the prominent German politician F.J. Strauss has provided a sensational commentary on his own trip to Moscow.

Among experts, no one denies that the Soviet Union has maneuvered itself into the economic cellar with its over-armament efforts, and that therefore, for this reason alone, *perestroika*—restructuring—would have been an imperative,

Brigadier General Paul-Albert Scherer (ret.) *was until 1977 the chief of the Militärischer Abschirmdienst (MAD), the Military Counterintelligence Service of the Federal Republic of Germany. He is presently a specialist security consultant in the F.R.G.*

▲ Speakers on the opening day's panel in Andover, Massachusetts, call for the replacement of the Bretton Woods System with a new, just world economic order. From left to right, Lyndon H. LaRouche, Jr., leading international economist, and U.S. presidential candidate; Helga Zepp-LaRouche, founder and chairman of the international Schiller Institute; Jorge Panay of Panama, who spoke on Panama's fight for sovereignty; Webster Tarpley, Schiller Institute President, United States.

◀ His excellency Dr. K.D. Kaunda, President of the Republic of Zambia and Chairman of the Organization of African Unity, sent a stirring message of support for the goals of the conference.

▶ Abdou Diouf, President of the Republic of Senegal, telegramed greetings to Andover and called for urgent action to solve Africa's debt crisis.

◀ Dr. Fred Wills, Conference Chair, former Trade and Foreign Minister of Guyana. Speaking before the United Nations General Assembly in September of 1975, he was the first government official in the world to propose moratorium as a pathway to solving the Third World debt crisis.

▶ Lyndon H. LaRouche, Jr. delivered the conference keynote, "The Tasks of Establishing an Equitable New International Monetary System," in which he proposed a detailed solution to the global economic crisis.

◀ Former Sen. Frank Moss (D-Id). Chief sponsor in the U.S. Senate of legislation to create the North American Water and Power Alliance.

◀ Farouk Shakweer, general secretary, African Association for the Promotion of Trade, detailed the effect of the debt pressures on perspectives for African trade.

▶ Mpinga Kalongi, representative of the Embassy of Zaire in Bonn, West Germany, spoke on the necessity of a new world economic order for Zaire's survival.

◀ Helga Zepp-LaRouche, founder and chairman of Schiller Institute, spoke at both Andover and Cologne conferences on "The Dignity of Man in a New World Economic Order."

◀ Irajá Rodrigues, Brazilian federal congressman, and Dercio Munhoz (below), economic advisor to PMBD (ruling party of Brazil), spoke on aspects of the Brazilian debt crisis.

▶ Mexican State Representative Patricio Estévez, of the PARM Party in Sonora, presented details on the economic crisis in his nation.

◄ Ricardo Veronesi, M.D., president, Brazilian society for Infectious Diseases, spoke on the international battle against AIDS.

▲ Fernando Quijano, the Schiller Institute's coordinator in Ibero-America, addressed the conference on the War on Drugs on that continent.

◄ James Frazer, Ph.D., gave an overview of the perspectives for science in a new world economic order.

American civil rights leaders addressed both conventions. Rosemarie Love (below), member of the powerful Cook County Commission in Illinois, spoke at the Cologne Conference on "A Vision for Mankind's Future." Amelia Robinson of Alabama (right), spoke at Andover on the "Fight for the Inalienable Rights of Man." Also at Andover, Rev. Wade Watts (above), NAACP President in Oklahoma, addressed the conference on "The Meaning of Economic Justice."

▶ Brig. Gen. (ret.) Paul-Albert Scherer, former head of the Military Intelligence Agency of West Germany (MAD), spoke at Andover and Cologne on the "Strategic Implications of Economic Policy."

◀ Donald Eret, former Nebraska State Legislator, spoke on "How International Pricing Systems Have Destroyed American Agriculture."

▶ Webster Tarpley, president of the Schiller Institute, whose presentation was titled "Who Is Responsible for the Coming Crash?"

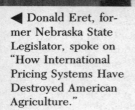

# Brig. Gen. Paul-Albert Scherer (ret.)

even years ago. There were, and still are, the gigantic costs for the five branches of the armed services:

1. Ever newer types of tanks, re-equipment of their land forces, following the negative experiences in Palestine and Egypt, with entirely new families of tanks, which was also an aggressive response to the introduction of the far superior Leopard II in the German Bundeswehr;

2. Many more squadrons for the air forces, new air bases, introduction of ever newer reconnaissance and fighter-interceptor, fighter-bomber models, and the development of a long-range air fleet;

3. In the context of their navy, an absolutely megalomaniacal ship construction program over twenty-five years, in order to finally overcome the traumatic inferiority complex in a nation with few ice-free ports, and foreign land barriers at the Baltic and Black Seas, a problem which became pronounced since 1904-05 (Japan's total naval victory over the Russian bluewater fleet at Port Arthur/Tsushima), and intended to overcome this inferiority in rivalry with the maritime power of the U.S.A.;

4. For the air defense forces, under the pressure of the geographic conditions of the largest country in the world, with immense air spaces facing China, the U.S.A., and Japan in Siberia, and in the West, a stationary and mobile air defense was built up, intended to compensate for the encirclement anxieties and potential compulsion of having to defend on 360 degrees of the compass;

5. The strategic rocket troops were developed and expanded with unprecedented expense of effort and resources following September 3, 1949, as the explosion of the first Soviet atom bomb demonstrated the most significant performance of Soviet espionage drastically to the entire world, and dispatched the American atomic monopoly into the historical past. At the same time, Khrushchov was boasting that the Soviet Union had developed the capability to hit a fly in outer space. The Communist Party of the Soviet Union undertook everything in its power to strengthen this "apple of the military-technology eye," with no regard for the living standards in the country.

If, in addition, we consider the immense costs swallowed

107

by the ambitious space program, because the first man in space just had to be a Russian, and the United States was supposed to be delivered the shock of being the rival who had been left behind—by the way, a typical example of Moscow's aggressive psychological warfare—then a little *red* light ought to go on in our minds. These costs were shouldered because Moscow, as Number One among the superpowers, wanted to have space stations in Earth orbit, which they could expand, and thus tangibly demonstrate the lead enjoyed by their own SDI program over that of the U.S.A.

It is also worth keeping in mind that, since 1966, the Soviets have been tinkering with killer-satellites, and have already reached the development and test phases, despite their "holier-than-thou" agitation about the United States.

Another element also deserves a place in this review of facts: the costs and efforts expended in the Soviet conduct of wars of all kinds in "peacetime," beneath the suicidal threshold of a nuclear war, wars going on now, right in front of our eyes, naturally without declarations of war. These include proxy wars, such as those which could be waged in Korea after the abolition of the U.S. nuclear monopoly from 1950 onward.

## Modern Irregular Warfare

There is unfortunately no really graphic and memorable, or correct, collective concept for these kinds and forms of war, conducted from the underground, and under employment of intellectual, psychological weapons effects, and on the other hand under massive employment of firearms, explosives, and other such instruments of war. Most of these phenomena can be captured in the term *modern irregular warfare*, as expressed by the distinguished author of the book by that name, which first appeared in 1972, and was recently republished in 1986, *Modern Irregular Warfare*, by Law Prof. Freiherr von der Heydte, an experienced wartime paratroop officer, and at the end of his military career, General of the Reserves of the Bundeswehr. Before this visit to the United States, I compared the views I am presenting here with his evaluation of the situation. There were hardly any differences in our views.

The preferred Soviet war in their striving for imperial dominance, to knock out the "decadent West" without having to resort to the big and heavy club of military weaponry, is the psychological-political war, with a psycho-cultural theater of war included. There can hardly be any doubt about the existence of this theater of war, because the majority of the expenses have to be paid daily in Western currencies. According to testimony by deserters, and according to the results of intelligence gathering, in the 1960s, before the "anti-Vietnam" and "Ho-Ho-Ho-Chi-Minh" student unrest, there were over 500,000 Soviet agents deployed for such agitation. With the installation of the "peace movement," which itself consists of 25,000 organizations, the number of these agents grew even more. The majority are agitators, manipulators with infiltration assignments, agents-of-influence. A minimal number of these are espionage and sabotage agents.

The stump of amputated Europe which was still free, and especially the Germans on the Western side of the Iron Curtain, in the zone where the weapons caches of East and West run up against each other, were the initial targets following the war, both in terms of area and personnel. These areas are the priority target once again: The aim feverishly pursued is to collapse these areas into neutrality. Immediately after the war, the "Without Me" movement against the rearmament of Germany was in the forefront, followed by the "Easter March" movement and the "anti-nuke" movement. These were succeeded by operations to disrupt the enactment of emergency laws, legal measures to remove limitations on job accessibility for communists in the civil service, and numerous instigations of strikes. The rate of agent training was hardly able to keep step with "requirements."

## Soviet Subversion Operations

Then, after the McCarthy phase, the U.S.A. itself became subversion target number one, steered out of the New York headquarters of the United Nations, as the grand residence of the Soviet intelligence services, in order not to cause any more talk about goings-on at the Soviet embassy in Washington. The assignment was destabilization via exertion of influence, and, of course, a continent–wide profiling of institutions of political decision-making, the think-tanks, the industrial laboratories, as well as the intelligence and counterintelligence agencies of the U.S.A.

The intelligence specialist knows that it takes years to build up an agent in the right position, to the point where he enjoys confidence, has achieved full insight into his target, and can thus influence or expose the most sensitive areas on behalf of his contractor. Controllers, the officers responsible for leading the agents, give their top agents up to thirty years of time to work their way to the very top through the establishment. At that point, they can become fully effective as "moles." Concretely, that means: For some twenty years, the "moles" have been working toward our demise in the United States as well as in Western Europe, and these are the most talented, the most foxy. They have reached leading positions, constitute factions within the apparatus of their respective target institutions, and exert their influence unhindered.

I am certain—just to point up one example—that many CIA blunders, with the deliberate retirement of many good officials from this instrument of security which functioned well for decades, are due to subversion of this sort. In addition, the early construction of Radio Free Europe in the area of Munich as a credible source of orientation for the Soviet population, despite all of the jamming transmitters, has much for which it must thank this agency. In comparison to secret services, it is of course much easier to go under cover and build a career in newspapers, radio, and television editorial boards, in the apparatus of government, political parties, trade unions, even in the armed forces, in universities, in scientific staffs, religious organizations, etc., and to exert efforts according to the assignments of the contractor and benefactor.

I do not by any means intend to promote agent hysteria, but rather to soberly demonstrate where the Soviet Union has brought us and itself with its challenge, its worldwide com-

## Brig. Gen. Paul-Albert Scherer (ret.)

munist underground activities, when, after victory in the war, and having prevented the promised free elections in Poland, Hungary, Romania, etc., on Stalin's orders, the Soviet Union embarked on an indoctrination for its missionary crusade.

Proclamations today that such a crusade does not exist do not mean it has stopped. We are still being subverted. One look at the peace movement, the "Green" parties, and at terrorism, is sufficient to induce us to be extremely cautious in our enjoyment of the 180-degree turn which Gorbachov has allegedly initiated, and to demand real evidence that this change ever happened.

## Is Moscow in Trouble?

Up to now there is only evidence that the former bread-basket of Europe cannot feed itself any longer, and that it would starve if there were a blockade against imports. Up to now there is only evidence that the economy cannot be managed in the same way any longer, because for years the necessary planning quotas have not been achieved in a wide range of production and supply facilities. The fact that the Russian economy lags behind Western productivity by ten to fifteen years is not solely due to Russian mass alcoholism. Up to now, there is only evidence that the West is being offered a form and a succession of disarmament steps, loudly and with ever-shifting proposals, which is to induce the West to give up the wrong weapons, at the wrong time, and at the wrong places to the detriment of Western security. "Do away with all nuclear weapons by the year 2000"—who would not want that, if it meant calm, real peace on this Earth? But it looks as though the weapons which neutalize the Soviet high-grade conventional superiority and hold it in check are the very ones which are supposed to disappear, thus allowing the classic conventional assault to be waged once again, with impunity, without entailing suicide.

The seductive argument for American ears, then, is: "Moscow's conventional strength would never be enough to carry the land war to North America." But this is certainly wrong.

111

## Development is the Name for Peace

The notion that a landing, such as that in Normandy in 1944, is no longer possible in the world of today, with forty times the firepower compared to World War II, simply misses the main point. Military people and politicians who think this way ought to know much more about Soviet *spetsnaz*, and ought to read the book referred to above on *Modern Irregular Warfare*—it has been translated into American English.[1]

It is here that we encounter the most primitive form of war that there is: the poor people's war, the desperate lunges of radicalized unemployed, the fanaticized of society who drew the all-too-short straws. Through instructor-agents, the leaders are easy to find in any country in the world, if a sufficient mass of conflict-explosives has accumulated. Communists always jump on a rolling train; they don't invent the railway to subversion. There are entire agent provocateur instruction programs with recommendations for producing rumors, ruthless use of lies, and deliberate deception, which are taught in Tashkent, Prague, or Moscow. The higher levels of instruction for agent controllers are offered by the Lumumba University. The successes are examined on the basis of new experience gathered in Germany, Angola, Great Britain, the Sudan, Ethiopia, Indonesia, Libya, Portugal, Italy, South Africa. The element of surprise in the conduct of warfare comes from out of the dark, suddenly, as deliberate and all-encompassing use of violence, with conspiratorial preparation, targeted subversion, conspiracy cells and agitational seeds in the population, where possible also in the armed forces. Again and again, armed fighters, without uniforms, appear in small units, on the model of the Spanish partisan war in 1808 against Napoleon, and they shoot, demolish, kidnap, take revenge, threaten, and conduct their combat in full military manner.

A current, large-scale example for this form of guerrilla war in "modern irregular warfare" is the Philippines today, with the strategic operations bases of the U.S.A. for the Pacific Ocean. Originally generated out of the resistance movement

1. *Modern Irregular Warfare in Defense Policy and as a Military Phenomenon*, by Prof. Friedrich August Frhr. von der Heydte, trans. by George Gregory, New Benjamin Franklin House, New York, 1986.

against the Japanese, the communist underground fighters on the 7,100 islands are aiming at overthrowing the government with Moscow's help, and then putsching their way to a coalition government in which they would determine policy. There are numbers of older examples in the Sandinistas, the Viet Minh, the Viet Cong, in Mozambique, and in Algeria.

# What the Soviets Gain in Afghanistan

That the guerrilla tactic can also result in successes in the West, is proven by the now eight-year-long combat of the resistance fighters in Afghanistan, although the help of the civilized world against this gruesome massacre conducted by the Soviets is clumsy, and incredibly anxious. Soviet imperialism is indeed suffering its military "Vietnam," in contrast to the political "Vietnam" of the U.S.A., but it is horrible enough to have to watch and see how millions of people in the world, as well as within the Soviet Union, have been sacrificed to this communist Moloch for seventy years. Yet, the talk is always about peace and disarmament, and now even about democratization. Words, big words.

The invasion in Afghanistan—this has to be said in the framework of the cost-benefit considerations—is surely an extraordinarily immense burden on the state budget of the Soviet Union, but for such a system, fundamentally oriented toward expansion, there are a number of important plus-points to balance the expenditures of effort and resources:

1. They are now only 800 kilometers from ports on the warm Indian Ocean, a strategic fact of the first magnitude, toward which even the Czars strove. A withdrawal from Afghanistan does not mean giving up a system absolutely toeing the Moscow line, which is the crucial point about the date of May 1988. Are we supposed to believe that what the Soviets succeeded in managing in Prague, Budapest, East Berlin, and in Warsaw, they will not succeed in achieving in Kabul? So-called "fraternal friendship," to be sure, comes about at the cost of freedom, but what power is there in the world which will prevent it from happening?

2. It is an additional and important strategic advantage, that Soviet armed forces can test and exercise with their weapons systems, with their own personnel, far away from the eyes of the world, including practical tests of newly developed systems, which would be impossible with Vietnamese or Egyptians. Hundreds of thousands of Soviet soldiers have gained combat experience, useful for all eventualities, and they have learned guerrilla tactics anew after a forty-year lapse in practical combat applications on a large scale.

3. The combat tactics of the *spetsnaz* special units can be tested in practice, and accommodated to the situation. This puts the Soviet leadership in the position to be able to deploy these forces at any time with guaranteed success, should war conditions in peacetime make their deployment necessary— these are military assault agents trained under permanent conditions of extreme hardship and the pressures of sustained combat sport, segregated in closed special camps with special language instruction for their country targets, drilled in techniques of silent assassination and demolition of command centers, as well as reconnaissance of strategically important information. The existence of over 30,000 such *spetsnaz* agents is known. They are almost always deployed in small units, but also have brigade staffs at their disposal. *Spetsnaz* is the acronym for *spetsialnogo naznacheniya*, i.e., troops for special deployment. It is understandable that their existence and purpose are kept secret, since Soviet deception propaganda increasingly emphasizes, that only defensive strategies are appropriate. Then follows the required demand of the peace movement for the Western armed forces to reduce their offensive weapons. That, of course, is a conscious total mobilization of the stupidity-potentials of Western publics, with the aim of outright cheating.

## Soviet Psychological Warfare

The more salient issue is the intention pursued by the employment of weapons, for the weapons themselves have no intentions at all. Weapons may be used to defend or to attack.

Defense, however, is inconceivable without mobility, and movement without armoring becomes nothing but a fatal sacrifice under the effect of modern weapons. The intent is to irritate the public, and induce guilt feelings among politicians when they allocate resources for weapons. In fact, the opposition parties in the parliament in the Federal Republic of Germany are in the process of coming to the view that the ostensible assault capability of NATO constitutes an obstacle to world peace. One can only shake one's head.

The continuous deluge with such and similar disinformation, delivered free of charge from Moscow, enables many citizens to have but a hazy capacity to perceive reality. Understandably, some judgments simply presume too much specialized knowledge. An immense specialized department of the KGB secret service invents an uninterrupted flow of new lies with scientific special staffs and the Central Committee department for evaluating the West's propensity for falling for the lies.

By and large, the employment of this psychological warfare poison leads to high casualty rates. The casualties are those among us with loss of perceptual capacities, people who look at things from stilted angles, who sometimes suffer from impaired vision or even partial blindness. The fatalities in these combat encounters and battles in this form of war live on as those among us who are totally blind, and saw away assiduously at the limb we are still sitting on. Despite these outrageous waves of nonsense, these years and years of outright cheating, we are still supposed to be so gullible and trusting, and simply act as if we can surely believe that these new gentlemen are far more solid than their predecessors. For the time being, as far as I am concerned, there is no security that anyone could buy a second-hand car from such people without a cartload of well-justified second thoughts!

When Khrushchov began the party reform in 1962, it hardly took two years, and the antagonized *nomenklatura* in the Politburo and the Central Committee drove him out of town. His name was struck out of Soviet history. Should not the West wait for clear proof that the Soviets are now renouncing the proliferation of their ideology? We will find out soon enough. And whether the Soviets are willing to give up military superiority, without any "ifs and buts," that will become evident enough when Soviet troop strengths have to be drawn down.

The West has twice literally fallen for the trap of protestations of good will, appeals for understanding, peace offers, promises of peaceful neighborliness; twice the West let the Russian bear rope itself onto the back of the West, and finally collapsed under the tons of weight of the untruth. Twice within thirty-five years.

The first time, when so-called peaceful coexistence was offered, our oh-so-insightful professional politicians took these offers at face value, because they thought this meant "live and let live, peacefully side by side" in the Western sense, without thinking of Lenin's instructions that for revolutionaries deception is a duty. It was in the era of "coexistence," that the missiles were shipped off to Cuba, that the wall right through Germany was constructed, and that the so-called wars of liberation in Africa were waged from Moscow.

The second grand plunge into the bear's trap, with a large deposit of trust and confidence on Moscow's account from the West, with foolish expectations in the aftermath for a revision, came over us like an avalanche, when, in the period of so-called détente, the Soviets' missile armament and dislocation program in the European part of the Soviet Union became ever more immense, and ultimately, despite pious declarations and signatures in front of the statesmen of this world in Helsinki in 1975, the Soviets promised to respect human rights, the right to self-determination, to guarantee free access, and then in December 1979 there began a Soviet illegal military invasion, and the murderous war against the previously free people of Afghanistan. As was also the case when the Soviets swallowed the European countries on their Western borders, the invasion was preceded by subversion of the government by the communist minority, with a putsch and an alleged call for help, the customary fraternal support. When the new head of government by the grace of Moscow began to criticize the war, he was summarily shot, and a new marionette was installed from Moscow.

It ought to be clear, after these bitter experiences, just

Brig. Gen. Paul-Albert Scherer (ret.)

where the trip with the Soviets in the boat leads. Our Western professional politicians, the people who think ahead of us, the opinion-makers and self-nominated opinion dictators, have helped us far too little, as the silent majority of well-behaved voters, loyal citizens, and willing taxpayers, to be able to recognize that we have a totally wrong notion of just what peace is. With the exception of a tiny minority, they themselves still have faulty ideas about what peace is, and what peace can only be, beyond the framework of family, groups, population layers, society, and state.

## Western High Civilization the Target

"A part of self-destruction lies in the lack of clarity in ideas," Socrates had recalled for his students in classical Athens during a critical analysis of Greek ideas about values, and he postulated as a principle of all action: "First of all, get the ideas clear!" Relating that to today: the Marxist-Leninists have been satanically masterful in the development of their own propagandistic world of ideas aimed at deception. They speak, for example, of "people's democracy," even though in their world there is not the slightest trace of rule by the people, but only a small clique of functionaries which sits on top of the people with police-state tactics and techniques, and squeezes it as far as possible. Communist systems live more from the deception of their adversaries than from the conviction of their followers. Contempt for human dignity, defamation, intimidation, and blackmail are among the features of the collective model of behavior within the system, but not less so toward the outside. "Exploitation" exists only where there are capitalists; only the party of the communists is "progressive." "Anti-communism" is the fundamental evil of our time, a far greater danger than communism with its anti-fascist world-view—all of this is argued, and some people still believe it.

Among the demonstrable facts of the attempt to annihilate Western high civilization without regard to costs and effort, there is the Soviet deployment to exploit gang-terrorism as a

117

specific form of modern irregular warfare in peacetime, aimed at achieving the destabilization of Western Europe. What the Soviet Union provides in this regard goes beyond training in East Germany and Czechoslovakia, and includes safe-housing against police searches, travel arrangements with covert re-insertion, and mediating contacts to terrorist centers in the Middle East. That is how the terrorist gangs in Germany, France, Italy, and Belgium emerged, with the aid and support of Soviet secret agents, with hard-core leadership cells, core gang-cells in totally isolated underground conditions, with support groups for supplies, procurement, and contacts, with covert safe-housing and protection-zones in the so-called sympathizers' swamp of the same fanatic mentality. There have been numerous murders, family tragedies, countless acts of sabotage, horrible assassinations, and economic losses running into the billions.

In the free part of Germany, a special form of street terrorism emerged with Soviet support, one which extensively exploited the too liberal laws on rights of demonstrators. So-called "Revolutionary Cells" and "Alternative" adversaries of the state deliberately stage situations where they engage in street combat with the police, occupy empty houses, and build them up into fortresses, or, with masked, large-scale gang-war troops, screened by allegedly peaceful demonstrators, conduct assaults on nuclear power facilities by storm, nuclear fuel reprocessing facilities, airport landing facilities in Frankfurt, etc. The casualties already number many dead and thousands of wounded. A special deployment unit of sabotage terrorists disrupts the energy supply grid by sabotaging of railway high-tension cables and cutting down or bombing the large metal electricity pylons. Years ago in Italy, the Communist book publisher and millionaire Feltrinelli, well known as a sophisticated promoter of the leftist subculture, was caught abreacting his contempt for society on such a pylon. His inexpert handling of the explosive charge caused his death.

Reviewing the countless number of Soviet nefarious activities, undermining normal Western expectations about freedom, individuality, law, security, happiness, and life, and evidencing a truly messianic form of crusader mentality with pathological traits, the inevitable question becomes that of why the financial bankruptcy of this megalomaniacal system did not occur long ago. Years of compulsory cheap supplies from the satellite states, once the Soviets had reached the banks of the Elbe River—Tito did not revolt without cause— ten years of reconstruction work done by German and Japanese prisoners of war, reparations paid by the West, annulment of war debts, continuous slave-labor of millions of prisoners in the labor camps and special prison camps, exports of gold, furs, and especially weapons, along with the most important of all from the standpoint of value—low living standards at the level of developing countries, and billions of dollars of espionage thievery of the most expensive Western know-how in production and research for over forty years.

It may well be, that the modern-day Lenins over there want to deceive us into believing they now want to make every effort to raise living standards, but are very hard pressed, even near bankruptcy, for all of the effort and resources devoted to expensive new developments in revolutionary new weapons, as the 1987 report of the American Secretary of Defense (*Soviet Military Power 1987*) suggests: Weapons to cause collapse of the nervous system over distances of more than 1 kilometer with radiofrequency/microwave weapons, while they view their own SS-20 missiles as obsolete, and throw them as bait onto the disarmament negotiating table. On the basis of the Soviets' observable mentality up to now, we should expect things of this sort from them. The most costly disinformation apparatus in human history up to now is still at work, its intensity unabated, and it still pretends to put the free world on trial. Of reform, real reform, there is nothing to be seen.

On this account, the astonishing observations of the Minister President of Bavaria, Franz Josef Strauss, formerly one of the most acute critics of the Soviet Union, are of little help, and carry little conviction. He gives credence to the sincerity of his discussion partners in Moscow, credence to their willingness to undertake a far-reaching disarmament, even to put their offensive posture of the postwar world to an end—this Strauss believes he can know from their words alone, nothing more.

That is tantamount to the recommendation that we ought to take the super-sophisticated television appearances of the elegant, attractive Raisa Gorbachova patting children on the head more seriously than her professional role as professor for aggressive Marxism-Leninism. Eastern psychological warfare nowadays does not present itself as a shoe-hammering at the speaker's pulpit of the United Nations. In 1984-85, communist bookshops in Germany had large posters with the letters: "Advertisement of an American travel agency: *Book your trip to Europe while Europe still exists!*" Nowadays people no longer want to be so stupid. Some American politicians, who want to pull the U.S. Army and the U.S. Air Force out of Germany and England, ought to know that Western Europe is the last bridgehead of freedom on the east Atlantic coast. Once the factories there have begun to work for the Soviets, the lights will go out in America—forever.

When we look at the economic and political situation in the world, we ought to recall Khruschev's words now and then, when at the beginning of the 1960s he said that the Soviet Union would catch up with and surpass the West. We suffer from a kind of continuous mental fog, which the Russians call psychological warfare. The Soviet Union has obviously entered a pact with not negligible parts of the self-nominated opinion-makers in the West, for it continuously succeeds in finding channels through which to spew out its disinformation. Subversion of editorial boards on the one hand, deployment of agents-of-influence in important groups, organizations, in the party spectrum and in areas of government, on the other hand, have produced a haziness of our political perceptions which reaches into at least partial blindness. By killing with silence certain positions opposed to them, citizens are supposed to slowly become incapable of exerting such necessary opposition.

Up to the Autumn of 1987, I belonged to the great silent majority of citizens in the Western world. But spurred onward and strengthened by the example of Lyndon LaRouche and his information, I understood that standing on the sidelines was the wrong way. Today one makes oneself guilty if one is silent. I was astonished that a private intelligence service can offer so much real and valuable information. These truths have to be spread among the people. I therefore want to appeal to you to do what I did in the Autumn of 1987, to take a position among your own acquaintances and no longer stay silent.

If we leave this conference as the bearers of up-to-date, hard information and positions, then over the long run we will break through the tactic of silence and black-out of the media. After the Second World War, I was a journalist and a chief-editor for ten years for a large daily newspaper in the Ruhr region, and I know that only the word from mouth to mouth can really fascinate and convince people. So we must build up a movement, in the sense of "to move something," out of the silent majority, we free citizens of an open society. Anyone who attempts to shove us off into the right-wing radical corner of the political spectrum is attempting, through character-assassination, to make uncomfortable ideas, independent thinking, and an awareness of the threat of communist psycho-terror methods permanently impossible.

DON LUIGI BOGLIOLO

# The Ethical Foundations of the New World Economic Order

"Today, it is clear that the simple accumulation of goods and services, even to the benefit of the majority, is not sufficient to create human happiness. . . . The experience of the last few years demonstrates that if the entire mass of resources and potentialities placed at man's disposal is not subject to a moral intent and an orientation toward reaching the true Good for humanity as a whole, then it will turn against mankind to oppress it. Equally instructive is a disconcerting statement made in the recent period: alongside the misery of underdevelopment, which cannot be tolerated, we find ourselves faced with a kind of overdevelopment, equally inadmissible, because, like the former, it is contrary to true good and happiness. This overdevelopment, in fact, consists in the excessive availability of all kinds of material goods. . . . [i]t easily enslaves man to 'possession' and immediate satisfaction, without any other horizon than the multiplication or continual substitution of things that one already possesses with other, more

Don Luigi Bogliolo, S.D.B., *is a dean emeritus of Urbaniana University in Rome, a consultant to the Vatican's Congregation of the Saints, which makes decisions on whether an individual should be beatified or sanctified, a renowned writer in the fields of philosophy and philosophical anthropology, and a member of the Pontifical College for the Study of St. Thomas Aquinas.*

perfect, things. This is the so-called consumer society, or 'consumerism.' " (John Paul II, *Sollicitudo Rei Socialis*, n. 28.)

To aggravate the situation, the world is divided into two economically (but not only economically) counterposed blocs: liberal capitalism and Marxist collectivism. This division is at the root of other, endless divisions. But division has always been a path leading to destruction and death; union and communion lead to life, peace, progress, and development. Faced with this very realistically described situation, the theme of the ethical foundations of a new, just world economic order seems to be very necessary. The historical reality that we are living through on the economic level, with its uncertainties, fears, ups and downs of the economic situation both in the East and in the West, calls for new foundations to be defined, solidly anchored to moral principles above and beyond the economic level.

## The Human Aspiration

The deepest aspirations of human nature are not to "have," but to "be." The aim of having is being, more specifically, the "being" of man who desires to be ever more authentically "man." Here we enter more decisively the matter of the ethical foundations of a new, just economic order, which is no longer sectorial, made up of divisions and counterpositions, but global, worldwide, above and beyond any ideological divisions. These ideological divisons, to keep themselves alive, are forced to allocate funds for unlimited military expenditures, which leads to the progressive impoverishment of peoples, nations, and individuals.

To clarify the ethical foundations of a New World Economic Order, today we have the document of a pontiff who knows the current world economic situation in all its aspects; he is very sensitive to individual rights and has specialized background in the matter as the former university professor of moral and social order. I believe, however, that it is necessary to dig more deeply: to go to the root of the key concept, to discover the very bases of the foundation of human ethics in all sectors.

I do not believe that I am erring in saying that on the threshold of the year 2000, we still do not know who man is. Doubtless, there are many human sciences which have made extraordinary progress in understanding the human body, its macro-structure and microscopic structure, through medicine, biology, and physiology. The study of single parts and single functions of the human body are at the root of specializations which take up the whole life of the doctor and surgeon.

But man is not only a body; we can say, he is not principally a body. Language itself among all peoples sees in man a duality of body and spirit. The corporeal component of this duality, as I said, has been studied and is continually being studied, to reach a deeper understanding; this is certainly worthy of praise. But the same cannot be said of the other component of this dual unity. Note: I am not saying, and will never say "dualism." Dualism and dual unity are not at all the same thing. Duality can exist without there being dualism. When dualism exists, there is a separation between two elements. Duality involves only distinction, but not separation. Who could separate intelligence from will! And yet, intelligence is not the same as will. Modern philosophical language often does not take this fact into consideration.

It may be observed that psychology has also made progress. It has said and does say many things about the human spirit. Yet, psychological studies (think of psychoanalysis, for example) study the human spirit at a phenomenological level. We are still at a purely empirical level, not very different from the biological-medical level. In fact, it is true that one tries to translate the functions of the human psyche into statistical data and numbers offered by machines and computers; no matter how sophisticated these machines may be, they are always the product of human intelligence. Therefore, intelligence is something more than the machines it has built.

Many philosophies are founded on human experience, in the realm of the senses, for example, empiricism, scientism, and more generally, positivism. It is not the case here to recall the historical origins of these philosophies, which are anything but realistic. They tend rather to mortify and impoverish the limits of human experience. In this sense they are damaging to experience, seriously reducing it.

This utterly partial, unilateral vision comes under the continual attack of scientific progress, which knows no limits. Human intelligence always reaches beyond all purely empirical data, thus, beyond any exclusively empirical experience. Reducing human experience to sense experience negates the most characteristic experience of man: his intellectual experience, which is as vast as existing reality. Human intelligence has no border other than "nothing," which can be neither a limit nor a border. It is characteristic of intelligence to grasp the existence of things; therefore, of all existing things, it knows at least one thing: that they exist, that they are, even though it is the task of science to clarify *how* they exist.

Therefore, today more than ever, it is urgently necessary to state what intelligence, as such, is. And it is the task of philosophy as a continuation and crowning of scientific knowledge.

## The Study of Human Intelligence

In our study of intelligence, its structure, and its functions, we are still in prehistory. In truth, some basic principles had already been glimpsed by the great Christian philosophers like Augustine and Thomas Aquinas, but they had no followers.

From then on, much rhetoric has been produced, but intelligence has not been studied in depth. Yet, man, or rather, the person, is the subject of moral and ethical rights, precisely by virtue of the fact that he is intelligent, and, as a necessary consequence, is free and has the power of will. Man is principally, though not exclusively, his *mind*, with the functions just mentioned. From this spiritual component springs his personal dignity, communicated fully to the body as well; thus, the being, its unity (the spirit is more intimate with the body and every part of it), humanization, the dignity of the person.

Penetrating the unsounded depths of the inner being of man, discovering its inexhaustible wonders, is extremely important. It is impossible to shed light on the ethical foundations of an economic order, or on any other sector of human activity, without the primary foundation of the person, which I call ontological, in the sense of the ultimate structures of man *qua* man. Ethical principles, or "that which must be," are solidly based on being, otherwise they disappear into nothingness.

Let us try now to identify certain characteristics of the spirit, or if you will, of the human mind.

1) Man is an indissoluble unity (for as long as he inhabits the earth), of body and spirit.

2) The spirit and mind have primacy over the body: A body without spirit is no longer man. Let us now note: The body is necessarily a subject, and included within the limits of space and time, tightly bound to the cosmos, in fact, a particle lost in the immensity of cosmic spaces. The spirit, with its intelligence, embraces and reaches beyond all time and space. It is no prisoner of time and space, of the cosmos. As a great professor of anthropology from the East acutely observed during a private conference, "Man as a body is really a microcosm, an atom among atoms; as spirit, man is the macrocosm which embraces and contains the entire cosmic universe lost in the intensive, not extensive, immensity of the spirit."

Each human being is the macrocosm with respect to the staggering quantitative dimensions of the cosmos. Each person is worth more than the entire sensible world.

From this it consequently follows that the human spirit transcends the body. To understand this, we may seek assistance from the great doctor of Hippo, Augustine, supreme philosopher and supreme theologian at the same time, when he articulates a principle which has always fascinated me: "All that which is better, is greater than any quantitative size." If the spirit is better than the body, it is greater than the body, if it is better than the land and water surface of the earth, then it is also greater, if it is better than the entirety of the physical universe, then it is also greater.

This means that the spirit has a superior existence in quality and intensity of density, incomparable to all that which is purely material. Man, therefore, insofar as he is spirit, transcends the cosmos. There exists not only the transcendence of God with respect to the entirety of material and spiritual creation; there is also the transcendency of man with respect to the physical cosmos. The human being is a paradox: It is both cosmic (as a body), spatial, temporal, historical, and (as spirit), meta-spatial, meta-temporal, meta-cosmic.

The staggering spatial dimensions of the universe itself provide an introductory pedagogy to the intensive, unsoundable depths of the spirit.

3) Here it is necessary to clarify the meaning of a term, which, though extraneous to science, is indipensable to philosophy. "To transcend" means to contain all the values of the transcended, in a higher, richer and different manner (in an eminent manner). In the world around us we find this law already at work: The higher beings—for example, the beings of the animal realm—contain in themselves the values of the vegetable world, and something more. One can no longer speak of superiority and transcendence without the premise of a higher and richer inclusion and presence of all the values of the transcended in the transcendent.

Faced with this extraordinary anthrolopological realism, Saint Thomas was not afraid to exclaim: "Man is, in a certain manner, the totality of being" (*homo est quodammodo totum ens*; Cfr. commentary to Aristotle's *De Anima* ch. XIII). "All of sensible creation has its synthesis in man" (Prologue to Book II of the *Sentenze* of Pier Lombardo). "In man, all things are contained"—thus, it is stated in the *Summa Teologica* (I, 96, 2).

The spirit's transcendence of the body implicates another consequence: "It is better to say that the soul contains the body than vice versa"—which fills one with respect for each and every man, even if not baptized, whom we see haloed by the halo of the spirit.

5) We are accustomed to consider the cognitive activity of intelligence as abstract labor, as passive receptivity of the ideas which things, objects impress upon our mind, through the senses. I would like to note here that cognitive intelligence is never simply receptive and speculative. The human intellect is the most intensely active energy in the world, even if it is as silent and unnoticed. To "know" means to identify the known object with oneself. If the known is of material nature,

inferior to the nature of the mind, it becomes elevated by the human mind, that is, spiritualized, transformed, re-created, and it acquires within the spirit the same mode of being of the spirit, that is, a mode of existence superior to the mode of being which material objects have in themselves. God is omnicreative; the spirit, by virtue of its intellective knowledge, becomes omni-re-creative. Ideas are like things identified with the spirit, which acquire the same (cognitive) nature of the spirit; it is the principle from which every human activity springs: aesthetics, technology, economics.

6) The physical universe stands in relation to the spirit as the part stands in relation to the whole. When, however, it is a question of interpersonal knowledge, then intellective knowledge cannot be elevating. Here the known and the knower find themselves on the same level; knowing, therefore, becomes reciprocal communion. When God is concerned, knowing is always anthropomorphizing.

7) We note another paradoxical characteristic of the person: Each person is something unique and irreducible, and at the same time, universally omni-open. Nothing and no one can be far from the spirit, neither spatially nor temporally. Each human being not only includes in an eminent and transcendent manner all values of the infrahuman, but is necessarily united, as spirit, to every other human being who was, who is, and who will be. Prior to every act of human solidarity, there is a natural solidarity inherent to the nature of the spirit itself. For the spirit as such, the essential, principal component of the human being, there can be no distances of space or time. It does not matter if man is only imperfectly aware of that sense of humanity that he sees in other human beings, almost as an extension of himself, an "other himself"; the expression comes from Saint Thomas, who does not hesitate to assert: "The entirety of men is to be considered as one man . . . different persons assigned to different tasks, are like the members of a single natural body. . . . Thus the entire multitude of men must be considered as one single 'collegium,' rather, as the body of a single man" (Cfr. the question *de Malo*, q.4,a,1).

Here we have inherent in human nature—insofar as it is

Don Luigi Bogliolo

spirit—itself, the overcoming both of closed individualism and of collectivism, which (due to its materialistic roots) ends up considering men "a mass," a "flock," a "herd."

Overcoming these two extremisms, which is of utmost importance for the purposes of defining ethical foundations for society and social economy in every dimension, is possible only when such foundations are built on the integrity of man and in respect for the constituent components of man, and leaving intact the past of the spirit. Everything is given to man, his very body and all material and cultural goods, for the development of his personality, which is capable of progressively perfecting itself throughout a lifetime. Only the man capable of growing, progresses in strengthening his own personality, progresses in mastering his own body, and mastering the use of material goods, his sentiments and passions. If he wishes, he can succeed in bringing to the inner person that equilibrium, peace, perennial serenity which, in the Christian conception, constitutes sanctity, of which even Kant speaks.

It is only from this internal peace that social peace derives, in all directions—familial, civil, national, and worldwide.

## The Rights of Man

And yet, the perfection of the human being, made of body and spirit, requires material goods.

When John Paul II in his latest social encyclical affirms resolutely that "the goods of creation are destined to all: What human industry produces from raw materials, with the contribution of labor, must be at service of all, equally," he is doing nothing but recalling a principle always present in the great Christian tradition, collected and summarized by Saint Thomas Aquinas, who said: "Temporal goods, conferred on man by God, are the property of him who possesses them, but their use is not only of the owner, it is also for others, who may be nourished by means of them" (*Summa Teologica*, II-IIae, q.32, a.5). Thus, the right to use of wordly goods is given immediately by God to all men. By virtue of the fact

129

that man is a person, he is, by definition, ontological proprietor of the world, with the duty to become a moral proprietor, which is something more than the right to private property. It is a doctrine deriving expressly from the Gospel, from the supreme esteem which the Gospel has of the person.

Here is the reason why a careful study of the dignity of man, even from a purely philosophical standpoint, illuminates his essential rights. And this applies not only in a strictly personal sense; it applies equally on the national, international, and world plane. The limits of private property extend to nations and the entire world, which has become one family today. A new, just world economic order must take these foundations into consideration. The human being, as such, is the immediate purpose of all forms of society. The laws of states and nations, even when they apparently limit personal liberty, have as their ultimate aim the well-being of the single person and the cohabitation of persons in different social levels.

The person is the primary ethical foundation of every ethical norm, be it individual, social, national, or worldwide. Economic goods cannot in any way be an end in themselves, but are always means for the sustenance and development of the person.

The important encyclical *Sollicitudo Rei Socialis* must be read in this light. The world economic order cannot be constructed ignoring the primary reality of man as a person; that is, implying the primacy of the inner to the outer, the primacy of spiritual goods over the entire infra-human world and over the goods deriving from it, naturally or through industry and human labor. It is urgent to devote attention to two constituent elements of human nature, to avoid founding the New World Economic Order on an abstraction. A partial consideration or concept of man (pure spiritualism or pure materialism) remains necessarily abstract and unilateral, and ends up generating an economic system which resolves economic problems exclusively on the basis of property rights, private or common, national or international. Such an abstract conception played a not-insignificant role in the origin of the conflicts that have bloodied human history, especially in Europe, and it is one

of the reasons why the world today is held in the grip of fear
of more serious conflicts, capable of annihilating humanity.
The confliction of hermetic egotisms must be overcome, if a
new era is to be inaugurated: The era of solidarity. A new,
more just, more efficient world economic order, the very
topical theme of this conference, is certainly of very great
importance.

Nowadays, the economic failure of one nation or group of
nations cannot occur without sooner or later having repercussions, directly or indirectly, on the world economy.

This explains the expressions that recur in almost every
paragraph of the new social encyclical: "interdependence" (nn.
9, 17, 19, 26, 38, 39, 45, etc.) is one such expression, which
is associated to the others: solidarity, collaboration, corespon-
sibility. These are concepts which the Pope specifies more
clearly when he expressly states: "We are all coresponsible
for each one and everyone" (n. 32) and also; "All of us are
responsible for everyone" (n. 38).

This springs from the relationship that binds each man to
every other man, as I outlined earlier. Just as one cannot go
against God without going against oneself, one cannot enter
into conflict with others without entering into conflict with
oneself.

The conclusions are logical: What is good for each is good
for all, what is bad for each is, in a certain sense, bad for all.

These foundations are to be taken into account when planning a new, more just economic order, especially nowadays,
when means of communication have practically eliminated
distances, and made the inhabitants of our planet members
of a single family.

# A Vision for Mankind's Future

I come to you from the great metropolis, located in the United States of America on the shores of one of the five great lakes, Lake Michigan. This great metropolis has one of the seven engineering miracles of the world, where the great Chicago River flows backward. This was done in order to protect the health of the city in its beginning days.

This metropolis, Chicago, is in Cook County and the state of Illinois. It's a dynamic mix of all that makes America what it is, and a true indicator of the direction that our country will take, either thrusting forward or backward, in the endeavor to champion the improved quality of human existence.

The mix referred to is the fact that Chicago is a transportation capital, including all phases; a manufacturing and technological center; an educational center, with two of the Ivy League schools; a health center with five medical schools; a trade center; and a center for securities and commodities exchange. Chicago is the headquarters for many national and international corporations. Chicago is the so-called melting pot of people leaving the vanishing agro-rural culture, into this urban malaise, whose process is in a continual state of synchronization.

The aura around all of this activity is the uncertainty of our future. The changes that will affect not only the people of

Rosemarie Love *is an elected member of the Cook County Commission, Illinois.*

Cook County, but the world as well, are part of this aura of uncertainty.

The questions I ask are: What repairs and what new infrastructure do we need to put in place to bring certainty to the world's future? What are the barriers? And now, what are our solutions?

We're slowly losing our place in America, because our nation is abandoning the monetary policies which produced what in the past had been our guides for national growth products, such as the steel industry, the auto industry, and the appliance industry, where our people could be placed into work and be assured of jobs, at even a minimum standard of income.

Now we have evidence of a vanishing working, blue-collar middle class. We are now finding young men twenty-five to twenty-six years old, who have never really held a job. We're finding in the transition to a service-type economy, ill-preparedness of the people who make up the orderly transition into those occupations of production. The major universities, such as Northwestern, Loyola, De Paul, University of Chicago, University of Illinois, the city junior colleges, and other state universities, lack the ability to teach the basic educational skills needed—skills such as reading with understanding; such as the ability to think. What the indoctrination in education is, can be another whole discussion, because we can only define what is produced in our schools by what comes out, which is almost a disgrace!

# The Threat of the Drug Culture

We find that 78 percent of the Cook County budget is devoted to the criminal justice system, leaving so many of the other human needs lacking. Couple this with the attitude so many legislators have, who look at cost first, rather than human development. And we see that in our elementary and secondary schools, there is a withdrawal of thinking from the administrators, about the existence of a wholesome family, cultural life, the need of our young people to be taught and educated in the cultural arts, both classical music and performing arts.

In the Chicago School System's Board of Education, there are only about eight teachers who are specialists in classical music and the arts! In developing our society, these are components which are necessary, in order to change what we call the negative aspects of existence.

I believe the demand for drugs is created by the failure to bring to our children the enrichment of classical culture. All this psychedelic non-music which is going around, is killing our children. We have to go back! We must go back! The drug czars have infiltrated our society, where the demand, and a depressed attitude of being against rather than for a positive solution are running rampant. We must be cognizant of this. And we must stop it at its source. Whatever it takes!

Just saying "no to drugs," is not enough. We've got to eliminate it. We have to eliminate the access and the desire for it. We, as nations have got to stop supporting the International Monetary Fund and the World Bank, which have enslaved nations, whose economies are strictly based on the drug trade, which is destroying our young, who are our future! Also, these same two culprits, the IMF and the World Bank, cause us to exploit, in collusion with many of our U.S. corporations, our own labor market in the United States and the world, by not allowing the creation of jobs for our own countries.

There is no such thing as the status quo! Therefore, there should not be any fear from the professionals who are part of the criminal justice system, in having the percentage of the budget of Cook County allocated to criminal justice reduced! It should be a downward spiral and not an upward spiral!

We've got to get away from budgets that eliminate human needs! I can't emphasize this enough, because that's what we are supposed to be about. We are supposed to be about creating quality human existence. The morals of our society have to be attended to—because we're all interdependent. The threat of AIDS is real! Tuberculosis is real! Other diseases, about which we don't hear much anymore, because they have not been conquered, are real!

The tragedy is that the technology is in place. We must

force the "wheels of action," and come up with those cures! In the 1940s, when the big tuberculosis epidemic hit, Cook County implemented basic public health measures, while engaging in a crash research program to discover a cure, and at the same time building a new hospital to handle the patients, in the event no cure was found for some time. Ironically, by the time the construction of the hospital was completed, a cure was found, and that hospital never housed one TB patient!

Today, with the economic and financial crisis, Chicago has over 40,000 homeless people, who have to suffer inclement weather and hunger, while we have bastions of storehouses filled with food for "imagined and planned emergencies." With the homeless, TB is on the rise again in Chicago.

## Plan for the Future

Unfortunately, today we have to start to think and plan for how we will care for the crisis now unfolding. Because of the reduction in the job market, which is causing the homelessness, along with the immorality that has been allowed to creep into our society, and the illegal use of narcotics, through the development of the narcotics culture, we are now confronting AIDS and these other diseases.

We have to make sure the chief obstacles to quality life, the IMF and the World Bank, stop destroying the positive gains that have been made and could be made still!

The idea of paying our farmers not to grow, when there is hunger and starvation on all continents of this world; the idea of two potatoes in Poland per week is despicable! The idea of no food at all in Africa; the idea that in a land of plenty like America there is malnutrition is an abomination!

The agents of genocide must be identified, and brought to trial, convicted and eliminated! We've got to put constraints on those who are fostering chemical warfare, as well, which includes the drug Agent Orange, and pollution of the water and airwaves.

Human existence must be protected. At the heads of our governments we need people in key positions, who have to think in terms of how we go about making sure in everyone's lifetime, people can enjoy the natural benefits of life!

We do have a spokesman on the horizon thinking in these terms. And that is Mr. Lyndon LaRouche. And he must be heard. Why is this concerted effort made to silence Mr. LaRouche? Individuals whose purpose is to inform and give an overview to the population are involved in trying to silence him. Who are we protecting?

Our Creator obviously intended for all peoples of the world to be a part of the world. Why should a collective few make the decision on what we should hear and what we should not hear?

Going into the twenty-first century, there is a realm of untold development, and the contacts we've already made in space, show us that that's the direction we should take! We can't hold back! We must be forever going forward! We cannot allow the status quo to hold us back!

Life can be made better here, on Earth. Just as Leonardo da Vinci projected the airplanes, the submarines and waterways, we must take hold of our venture to space. Why, I even remember the comic books I read of Flash Gordon and space travel. This was a child's readings and imaginings, but these things came to pass in my lifetime.

In the future, we must not be naive. There will be settlements in outer space, and we must be prepared to go there as the Good and not bring the turmoil and human suffering, that we have all created unconsciously and accidentally, and, by that "privileged few," directly.

The United States must have a Democratic President for the next four years, with the ideas of Mr. LaRouche: one who has not been corrupted by the sins and greed for personal and individual gain; but one who has a sincere, forthright thought process for the continual existence of mankind!

# The Fight for the Inalienable Rights of Man

I wish to take this time to recommend to you what's going on, and the part that we have to play, because we are really the vehicle upon which we are going to save this country. I also would like to commend each and every one of you for being here, whether you're from the east, the west, the north, the south, or wherever you may be from, throughout the entire world.

We realize also, that we are in a crisis. If we do not do anything about it, the world will go under, and, of course, we will be with it.

It takes people like you to restore the integrity, the accountability, to the lawmakers of this and other countries, through your wise understanding and hard work, to right the wrongs. The right leadership, with whom we are working, you may rest assured, will replace the evil with good, and hate with love, thus paving the way for a democracy.

The world is on the brink of despair, and not only despair, of destruction. We realize that it needs leadership. It is like a crippled ship, riding the angry sea in high seas, with a captain

Amelia Boynton Robinson *is a leader of the civil rights movement in the United States, having marched and worked directly with Dr. Martin Luther King, Jr. in Alabama, where she continues that work today with the Tuskegee Institute.*

who is asleep, and its mates as drunken sailors. We are drifting. But there is still a little time left to reconsider and to regroup, to recognize and get back to the specifics. The World Bank's policies, treaty agreements, and all of the things that are going on now need to be straightened out.

Black America knows what is happening to this world. We know about the economic pressure, for we have been a part of it. We are the ones who can tell, because of the fact that we have experienced it. We know that there are many people who have ridden on the backs of the poor and the oppressed.

They have used also the method of pitting blacks against blacks, in order that they may annihilate each other. They have caused black people to forget their dignity, and their self-respect, by the news media putting them on tv and radio whenever there are murders, wherever there are drugs, wherever there is immorality. All of this is done in order that blacks and hispanics can be kept down, and feel inferior.

When Martin Luther King, Jr. came, he disturbed this kind of goings-on, and he began to make people feel as though you are what you contribute to this world. Yes, Martin distributed his ideas, and he distributed the meaning of what people are, and what they should do, because we are our brothers' keepers.

When we were struggling in Alabama and Selma, we found that we were not the only ones who were being mistreated, being used and abused, accused of things that we had not done, and excused from our jobs and from trying to be what this country stood for. In spite of that, some of us—like when the sailor goes out to catch fish, and some of them get out of the net—some of us decided that we are part of this world, and a part of our country; that we will contribute everything we can, to make the world what it should be.

Consequently, we have people who have made great contributions, but they have not been given credit. How many people of the world, and how many black people realize that a black man was the one who made the blueprints to take the bugs out of the lights and the telephones. And how many know that a black man designed the type of machinery that created the type of shoes we are wearing now. They don't tell us that.

They don't tell us that William Hale was the first to operate in open surgery, and save a man's life. They don't tell black people that the traffic light, that has saved many lives, was invented by a black man. They don't tell us that.

They don't want me to feel that I am somebody. But you give people a chance. You let them know that they are somebody, you let them know that they have potentials, and help them to create those potentials to the best of their ability. Then there will not be as much money spent for jails, and prisons. I understand it's around $100,000 for four years in prison, and an average of $28,000 that is appropriated to educate one person. We could close these prisons, and use the money to help to make people make contributions to this country.

When the news media brings forth the "evils" of hispanics and blacks, and when people believe it, we have created a monster, that we have to deal with. We "deal" with it by putting a whole lot of money in prisons, by "fighting crime," and by giving a hand-out in the way of welfare. We have to do something about this.

America is great. We need somebody at the ship's helm in order that they may be able to bring forth out of America all the good that this country has. We have been the kind of country that other countries have made an attempt to follow. We have great natural resources. We have been a leader for many nations. We have been a melting pot, for this country is built by people from all over the world, yet no country can claim her, for she is a part of many countries. She has been able to start from the beginning, having growing pains, and she has had labor pains. And when she became fully grown, I think she realized, "now I have made it," and began to slip back.

However, when we think of our Constitution, with the exception of the Bible, the wise men of this country got together and wrote one of the greatest pieces of literature that ever had been written. Our Constitution is the guideline for

America and for other countries. Having faith in God, they perpetuated this by putting on our paper money and our coins, "in God we trust," and "liberty." This is our Constitution. It guarantees the inalienable rights of man, politically, equally, and all men are free, and created equal.

What happened to the sacred vows that these great men wrote? Are we abiding by our Constitution? Do we give every American a chance to make his contribution, without being biased? Have we been fair to our farmers, our black, red, brown and yellow brothers? Have we been fair to our white brothers? Who holds in their hands the key to unlock the door of human justice, non-violence, equality, and love for all man? Who holds that key? Only those who have pure hearts and who believe that all people are created equal.

Have we become a part of the problem, or a part of the solution? We have a lot to accomplish, but it can only be accomplished by cooperation. We must fight against evil forever. We must see that the evil forces do not take over.

We realize that we are in a crisis, and that there are things that we must combat, such as AIDS, genocide, and a number of other things. We must make a sound recovery, in order that we can again take our place in this world.

Is it fair for Third World starving countries to have to send their produce to our country, in order that they may pay their debts? The greatest labor of love could be shown, if we showed our interest and humanity for all people, and if we sent technical and scientific assistance, into Africa, for example, in order that we might be able to bring running water lines across the continent, and help Africa to again feed herself. Give a man a fish, and he will be able to eat well for a day. But, if we teach him how to fish, he will be able to eat the rest of his life, and water across Africa will do just that.

America cannot continue to burden these small countries the world over. It is better to sit at the bargaining table, work out a plan to reschedule payments of our debtors, and work together in peace and harmony. I firmly believe that this method is better than the big-stick method, which is to divide and conquer.

The developed nations have an obligation to the deprived

and poor people, and even of this country, as well as others. We have to stop using and abusing people.

We cannot take the culture away from people, but what we want to do is to be able to develop the very best in every man. This can be done by sacrificial development of the under-developed world, and of Americans right here in this country. We want to improve the standard of living for all peoples, wherever they are.

The crisis we are going through strikes at the very heart of American citizens. The rich are frustrated because of the stock market crash. The farmers are committing suicide, and the poor are being reduced to crime and drugs. The IMF has become a conversation piece. I challenge America to begin now to fight for the very uplifting of this country and other countries, and to be able to put everybody on a standard of "I am somebody." There are greater things to be done than to bicker about little things. We have to use wisdom, and use our own minds, rather than that of the news media. I think we realize that the news media has been one that has actually destroyed dignity and self-respect in many people. The news media has not only done that, but it has used the public news-casting everywhere, in order that they may tell you the things that perhaps you would like to know, and not what is true. The news media has not been fair to the people of the world, nor the people of America.

The news media has really taken away from our country, from our people, the right to think for themselves, the right to believe what is really going on and what is true. Because most people believe just what they read, and if you say to them (and I'm quite sure you have found this out) "do you know such and such a thing has happened," they will tell you "no, the news media has said thus and so." I have asked a lot of people to tell me what they think of Schiller Institute, and of LaRouche, and what have they told me? The news media says this and that. I said, "Isn't it time for you to believe yourselves?" This is one of the things that Martin Luther King tried to get Americans to do. To believe in yourself. And believe that you have the same type of mind that thinks. And if you read through something, think through it to see whether it even sounds reasonable and sensible. But as a rule, we just believe what we read, and we stop with that.

We have a man whom I feel is a man of God. Just as Martin Luther King came to this world to save this country, I believe that this is another leader, who has been designated by God, because of the fact that he is humble, he is understanding, he is wise, and if we do not do something to save this country and the world, I fear, we will all go down the drain.

Whom shall we elect for our President and our leader? It's up to you. Like Martin Luther King, I think Lyndon LaRouche is a God-sent man. I think Lyndon LaRouche has not decided within himself that he wants to be President. I think he has an inward force, that is saying, "things have gone too far, this world is about to topple over, and I have been given by the Supreme Being the tools, the mental tools, to help to straighten this country out."

I believe he is a Moses, designed by the Supreme Court and by the Supreme Being, to lead this country and lead the world out of the chaos it is in. He is the one who is to lead us from darkness into light, because he is humble enough to listen to suggestions; strong enough to lead this country and the world to self-sustenance; and kind enough to have compassion for the less fortunate, instilling in them that they are somebody. He has wisdom enough to devise a program so that each person, who is normal, can exhibit his great potentials. Mr. Lyndon LaRouche will be the kind of President who will not deceive us, nor brainwash us about the happenings in the news media and what is going on.

He believes in the Preamble to the Constitution of the United States of America, as a guideline. He shall secure the blessings of liberty to all, because he believes that all people are created equal. He believes that we, the people of the United States, in order to form a more perfect union, to establish justice, ensure domestic tranquility, provide for the common defense, promote the general welfare of mankind, and all this, we will return to the standard for which this country was born.

Then, and not until then, will we lift up our voices and sing, with a clear voice and a clear conscience, my country 'tis of thee, sweet land of liberty, of thee we sing. We shall overcome the hatred, the obstacles, the thorns and the thistles, which our enemies have hurled against us and have put into our path.

With the right type of leadership, long will our land be bright with freedom's holy light. Then, and not until then, will we truly be able to sing "Free at last, free at last, thank God Almighty, I'm free at last."

# The Meaning of Economic Justice

To Mr. and Mrs. LaRouche and my Heavenly Father's children: You know, some way, some how, I have developed from a pessimist into an optimist. They tell me that a pessimist is a man who goes around wearing a belt and suspenders, and is afraid he's going to lose his trousers. An optimist doesn't wear a belt or suspenders, and doesn't care if he does lose his trousers.

I want to say this morning to some of the people from foreign lands, that there are people in America who care about you. There are people in America who are concerned about what is going on in your countries. I couldn't be happy with what's going on in Afghanistan. You've got people not happy with what's going on in Poland, not happy with what's going on in South Africa, in Nicaragua, El Salvador, and around the world. We are concerned and want you to know that we love all of you.

I want to say thank you to Mr. LaRouche, who, in 1985 on Martin Luther King's birthday, put 600 busloads of people in Washington, D.C., and then the enemy said that there were 20,000, and if you looked at the crowd, it was a whole lot closer to 40,000 than 20,000. But they minimize us whenever they can.

Yesterday, one of my friends from Oklahoma made a beau-

Reverend Wade Watts *is President of the National Association of Colored People (NAACP) in Oklahoma.*

tiful talk here. He said he was a Republican, and I'm here today, and I'm a Democrat. That makes a pretty good cross-section. And to tell you the truth, I might have been a Republican myself, if it hadn't have been for a fellow you might have heard about named Herbert Hoover.

I grew up on the farm and we were so poor, that the poor people called us poor. I remember we had corn bread and molasses for breakfast. We had molasses and corn bread for dinner, and had both of them for supper. And you could go to school, and if your mouth wasn't greasy, the little kids picked at you, because they knew you didn't have meat for breakfast. Momma had an old skin, and when we'd get ready to go to school—there were six of us—she'd call us and wipe our mouths with that old skin, so the little kids wouldn't pick at us and think that we had some meat for breakfast.

I remember, I left home about that time, trying to make it better. I caught a freight train and I went to California. I didn't have but $1.35 when I left, and when I got to California I was broke. I met a white fellow and I asked him for money to get something to eat. He said, "Son, I'm almost in your shape. I've got no money. But they've got a soup line down on 16th Street," and he said you can get some soup. I went down and got in that soupline about 9:00 o'clock, and I stayed in it 'til fifteen minutes to five. Three people before they got to me, they ran out of soup.

I sat down and wrote my dad a letter. I told him the condition I was in: my shoes were worn out, my clothes were raggedy, I hadn't eaten anything in two days, and I signed off, "No mon, no fun, your son." So in a few days, I got a letter back from the old man, and he writes me a pitiful letter. He said the old milk cow had died, and one of the mules was crippled, and he signed off, "Too bad, too sad, your Dad." But, he said, "son, if you can make it home, we'll kill the fattening hog." And I caught a freight train, and in a few days I was home, and we went out to kill the fattening hog, and he was so nigh between the eyes we couldn't shoot him, we had to smother him to death.

I've heard them talking about the bankers. That's something else I know something about. We had a farm, with 25 acres.

We rented 100 acres. Back then, you could borrow $150 to make a crop. And I saw them, back in the '30s, clean my dad completely out. They drove twenty-six head of cow out of our lot. Took two teams of mules, got the harness, took all the chickens and hogs, and Momma, standing there with tears in her eyes. I know what the bankers will do to you.

My brother bought a farm and lost it. Two or three years later, he got money to buy another farm, and a fellow came to him and told him, "I'll tell you what I'll do, I'll give you a deed to it." My brother said, "I don't want a deed. I had one with a deed and the mortgage took it. I want one with a mortgage. A mortgage is more powerful than a deed."

I've got a friend that I came here with, and I want him to stand up. George Gentry had 1,500 acres of land, tractors, combines, and everything else. They cleaned him out. Just they did to us back in the '30s.

We had a lawyer. During the last election I was talking with him, and he was asking me whom I was going to vote for, and I said, "if Lyndon LaRouche is on the ballot, that's who I'm going to vote for." He said, "I'm going to vote for President Reagan." I looked at him, and I said, "why?" He said, "I've processed more bankruptcies than I ever have in my life. And I get $700 apiece for them."

## National Insanity

They say that President Reagan said, when he became President, that he had never visited an insane asylum and he wanted to go to one. So they carried him out there. And the management told him, "now, Mr. President, we've got four men in here who believe they're Julius Caesar. We've got two who think they're Hitler. We've got six in here who believe they're Napoleon Bonaparte. Incidentally, we've got fourteen in here who think they're you! They think they're President of the United States." He said, "well, turn me in there with them. Those are the ones I want to visit with." And they said he and those other Presidents laughed and talked and had a big time. And finally, the manager said, "Mr. President, visiting hours are up." And all fifteen of them made a break for the door, and one of them got out, and the one that got out is giving this country one hell of a shame.

The other day, we had Mr. Gorbachov. They're about ready to change leaders over there. Whenever I learn how to pronounce one of their names, they're ready to change. Mr. Reagan was bragging about how he got the best of Mr. Gorbachov in the missile treaty. Now, I want you to watch this. If you had ten missiles, and I had 100, and I say that I'm going to destroy six every time you destroy one. Well, you've got ten, you're going to destroy nine of yours, I destroy 54 of mine. That still leaves me with 46 missiles and you with one. This is that great deal that the President beat Gorbachov on.

Congress said that they were going to give the people on food stamps a 68 cents-per-month raise on their food stamps. And the President of the United States said, "if they do it, I'll veto it." That's what he thinks about poor people.

Hulan Jack, in his lifetime, when they started to bring the dope in here, they were shooting it into Harlem. Jack said he'd introduce a bill in the [New York] General Assembly to crack down on the dope peddlers. They killed the bill. Adam Clayton Powell, who was a congressman at that time, he introduced a bill in Congress. They killed it. Now dope is everywhere.

A boy was given thirty-five years for one marijuana cigarette. At Ardmore, 115 miles from there, they caught an airplane load and couldn't convict anybody. This is America. This is justice, American-style. When you've got money, you do just about like you want to, in this country. You get all of the justice that money can pay for.

We've got another deal. I've heard Mr. LaRouche talk about this, called the Club of Rome. This fellow Cecil Rhodes and the Harrimans and that bunch have rigged up a scheme that's still working, to depopulate the black, brown, and yellow races by 85 percent. This is what they want to do. And then they said "it's overcrowded, it's overpopulated." Africa, they say, is overpopulated. It's one of the most underpopulated areas on the face of the earth.

The way the Club of Rome intends to depopulate the earth

is just like they're doing in Ethiopia. Pull that International Monetary Fund out and let them starve to death. Then you can't accuse anybody.

## FBI: The American Gestapo

We've got in this country what they call the "gestapo." Some of you have heard about that. But they don't call it "gestapo" here, they've renamed it, they call it the FBI. You know, there are a number of us, and especially in the organization I represent, who don't believe that Lee Harvey Oswald killed President Kennedy, and we don't believe James Earl Ray killed Martin Luther King. We believe that there was more power behind this.

Whenever you go to do something good, somebody is always ready to attack. I want to throw out just a few questions. And I want you to ponder over this. Who was it that offered a million dollars to have Martin Luther King killed? Where did James Earl Ray get the $10,000 he spent between the murder and the capture? Why couldn't District Attorney Jim Garrison from New Orleans get James Earl Ray's father to testify in the assassination of President Kennedy? Mr. Garrison was getting so close to that thing, that they had to get him to shut up. I believe that if he hadn't shut up, they'd have killed him.

Another question I have, is why did James Earl Ray go to Birmingham and buy a rifle, then the next day go back there and exchange it for another rifle? If he was going to use a rifle, don't you think that he'd have sense enough to know what kind of rifle he wanted? When he brought this rifle back, someone said "yeah, that's just the kind of rifle we wanted, lay it on the seat here." James Earl Ray's finger prints were all over it. Whoever handled the rifle from then on had gloves on. When they made the shot, they dropped the rifle, and when they found it, James Earl Ray's finger prints were on it, and they had put him on the move. I haven't got the time to tell you all about how the thing went, but you would see my point if I had the time to go through it.

[On the day of his assassination when he checked into his

148

Memphis motel room] how was it that Martin Luther King got room 306M, the only room that's in direct counterposition to the old flophouse across the street? I asked them at the Lorraine Motel, "How did that happen?" They said, "Reverend Watts, the city came down here that morning and cleared a path through there to room 306." Don't you know that I've got sense enough to know, that somebody in high government had something to do with that?

They had two police officers who [were assigned to] stay on duty till 6 o'clock. They pulled them off at 4 o,clock. They were getting it over the police cars that the killer went west, when all the time he had gone east. This is some of the shady work that goes on, that makes us highly suspicious of our own government.

Let me tell you another operation. When Mr. [J. Edgar] Hoover was in there [as FBI director], you could run for the U.S. Senate, and it looked like you were a good prospect. He'd call you up and tell you that you're going to need some protection, and I'm going to send a couple of FBI agents down there to protect you. What he's doing is sending those FBI agents down there to get everything on you they can find. Then, when you get in the Senate and you don't go his way, he comes up with your record. "Here it is. I'll expose you. I'll do this to you." And you could have told me, when I was a boy growing up, something bad about J. Edgar Hoover, and I'd have been ready to fight you. And then I came to find out that he lived his whole life with another man!

Here's what Moses said about it. This is in the 20th Chapter of Leviticus, 13th verse. "Mankind shall not lie with mankind as with womankind. It is an abomination before the Lord. He shall be put to death." Now Moses killed them back there. I don't say it was right, but I can tell you one thing. We didn't have AIDS then.

## Corruption in High Places

The FBI will take a little case, blow it way up, and the big money boys, who are really getting away with the money,

149

they never say a word. We had a little case in Oklahoma, where the school principal was making the children work on the CETA program and giving them half of their money. We were forever trying to get something done about it and they jumped on our investigator and nearly beat him to death. To this day, we've never been able to file charges. They won't accept charges. The FBI knew about it, the Justice Department knew about it, our U.S. senator knew about it, President Reagan knew about it. Now you ought to know what kind of country you're living in here.

Whenever you start out to correct a situation, to stop the corruption in government, it is extremely dangerous. Martin Luther King, as long as he was leading a few marches and preaching a few sermons, they didn't like him, but they didn't kill him. But when he talked about closing down the Vietnam War, he was talking about stopping the du Pont's from making that smokeless powder, he was stopping Remington Arms from making those rifles, stopping General Motors making those tanks and trucks, stopping the clothes manufacturers, so they had to get rid of him.

The Apostle Paul—if you remember, Paul was doing all right. They put him in jail and he was praying his way out. But when he messed with Alexander the silversmith, and Alexander had gotten extremely rich selling idol gods—he'd come to your house and tell you what a god he's got, you set him up and he'll cure all your ills, and the people were buying that stuff. Paul tore into it and was telling the people this is no god at all. You know what happened to Paul? He wound up with his head under the chopping block. They cut it off.

Jesus did pretty good, as long as he was healing a few people, making the blind see, preaching a few sermons. But then he went up to the temple, and up there they had money changers. You couldn't spend any money in the city, unless you changed it into temple money. You go in there with a $100 bill, they cut $50 out but they give you $50 back. You could spend that $50. If you wanted to offer a sacrifice like everybody else, that would cost you $2. A dove, that would cost you $5. Jesus looked at that and he said, "ye have made my Father's house a den of thieves, when it should be called a house of prayer," and made himself a whip and drove them out. They just went right on out and went to figuring out how to get rid of him. He lived eight more days.

The point I'm getting at is, when you go to digging into the corruption of people in high places, they're going to try to figure out a way to get rid of you. They kept putting pressure on Jim Garrison in New Orleans. If they hadn't, he was going to dig up the man that killed President Kennedy. This is one reason they had to get rid of Bobby [Kennedy]. Bobby would have eventually come up with the man who killed President Kennedy.

I was at Mr. LaRouche's house. I stayed till 3 or 4 o'clock in the morning. I listened to him talk about the situation in Africa. He is the only man that I have heard that came up with a genuine plan for Africa. Most of the people that go to Africa, go there to exploit Africa. They don't go there to help. I heard Mr. LaRouche talk about trapping the water when it falls, building lakes, tunnelling through a mountain to bring water through from the other side, getting tractors over there, trying with oxen, getting a fertilizing program. He's got a wonderful program for Africa.

But to be in your position, Mr. LaRouche, you've got to have the faith of Abraham, the courage of the Apostle Paul, and a willingness to suffer as Jesus did. I have read your literature, I have heard you talk. I have studied you. If there is one man that I'd like to pattern my life after, it would be Lyndon LaRouche.

That only goes to show you that the all-seeing eye of the great God is always looking through us and in us with the eyes of our fellow man. Every deed, regardless of how small, is being recorded by that great stenographer, who cannot forget to dot an "i" or cross a "t." So exact that every hair on our head has a number. So observant that when he comes again, the award he'll bring will be made out in full. If that should not suffice, he states plainly that "my grace is sufficient." Fight on, Mr. LaRouche, fight on! Fight on until the sun that drives the day breaks out of the stable over the universe and goes galloping headless from the east to the west its last time. Fight on until Abraham comes back, and gives a mathematical calculation of the seige that's already up in glory. Fight on, till Daniel comes back, rolling his stone that was hewed out of the mountain. Fight on, till old man Ezekiel comes back, and gives us the address of the blacksmith that laid the wheel in the middle of the wheat.

# The War on Drugs

If we want to speak about the dignity of man and the New World Economic Order, I feel it is very appropriate now that we focus our attention upon the war on drugs. As Pope John Paul II has stated many times, and others previous to him have also stated, man is the subject of economics and politics, not the object. And in the drug world, or in narcotics trafficking, we see quite the opposite, and we actually see the best or the worst expression of the things we have been talking about here. That is, of a Russian empire and of the usurers of the International Monetary Fund.

The focus of the war on drugs today or the non-war on drugs today, is Colombia. In Colombia, we have the situation where the drug mafia has openly and formally declared war on the state and all of its different instruments. Last week, as most of you know, they tried and assassinated the attorney general as a traitor because he insisted on the extradition of drug traffickers to the United States.

There was formally a treaty of extradition agreed to by both countries and implemented into law in 1979.

We should look at who these drug traffickers are, and what their apparatus is. *Fortune* and *Forbes* late last year had a list of the twenty richest men in the world. Among the top fifteen of both lists were Pablo Escobar Gaviria and Jorge Luis Ochoa,

Fernando Quijano *is coordinator of the Schiller Institute's work in Ibero-America.*

both from Medellín. These are the two top people in the Medellín Cartel.

Also among the drug traffickers are several other key families, all from Colombia. Cocaine, or the coca leaf is produced mainly in Bolivia and Peru—Peru being the largest producer of coca leaves in the world. It is transported to Colombia where it is refined and then shipped to the United States. So, Colombia is the industrial part of the operation.

In the United States, it is distributed not by the Colombian mafia, but by the Italian-American mafia that exists throughout the United States. Through them it is shipped to different ethnic mafias, the Chinese, Cuban, etc. That's how the process works.

Four hundred thousand of Bolivia's population participates directly in the coca growing, in the first phase of refinement and transportation. That's 400,000—almost 10 percent of the Bolivian population. Bolivia has a population of about 5 million—probably 6 million, if you count the anthropologists that are studying the Indian populations. It is one of the small Latin American countries in population, but in size it is bigger than Texas and California put together.

As you go upwards to Peru—the largest producer of coca leaves in the world—and it's been estimated that approximately $10 to $12 million is involved in the Peruvian part of the operation per year. Then, you go into Colombia, and that's where the industrialization or commercialization of the trade is achieved. This, of course, is no great achievement for Colombia. This is not really industrialization.

The people who have trained the Colombians in this task should also be known to people here. For example, Meyer Lansky personally trained several of the top money experts of the Medellín Cartel in the art of money-laundering. Robert Vesco, the man who essentially took over from Lansky as the U.S. mafia's expert on money-laundering, and moving funds from one place to the other—the financier of the mafia—is a partner of Colombian drug trafficker Carlos Lehder, who is now in a U.S. prison, who was the last person to be extradited from Colombia. They were business partners. Robert Vesco lived in Costa Rica for quite some time. Then, when it finally got too scandalous, he moved to Nicaragua under the Sandinista regime, lived there for several months, and it became so controversial that he then had to move to Cuba, where he now lives and is protected by Fidel Castro. We made the charge that he was living in Cuba in [the book] *Dope, Inc.*

Finally, Castro had to admit it in 1985, and said that Vesco was a good family man—he didn't say which family of the mafia—who was being persecuted by the U.S. intelligence agencies. And that he, of course, would defend him.

Carlos Lehder has openly stated in interviews that Robert Vesco was the man who trained him in money-laundering and was in charge of the financial aspects of the drug trafficking. Drug trafficking is a $500 billion a year enterprise. We estimated that figure about two-and-a-half years ago. People scoffed when we mentioned that figure then, but it is now widely accepted as the figure that the Drug Enforcement Administration and others use.

That was two-and-a-half years ago, it's probably more now. There are 20 million users of drugs in the U.S.—regular consumers of drugs. That's almost 10 percent of the population. It is stated that 60 million in the United States have experimented at one time or another with drugs.

This is, however, not only a problem in the U.S. It has reached Europe in huge proportions. And even in the producing countries—in fact, as we know from the recent kidnapping of the mayoral candidate of Bogota, Andreas Pastrana, the biggest problem in Bogota, a city of over five million, is drug addiction.

So, in the producing countries, one of the biggest problems among the youth—and that includes Peru—is drug addiction. And it has swept the countries like wildfire.

This gives you a sense of what's involved here in terms of the mafia. They are very powerful because they have now—with the acquiescence of the political class of Colombia—forced the government to stop extradition of mafia leaders.

And here is where you begin to see the collusion between the Russian empire and the IMF with the drug mafia. First of all, the extradition treaty was challenged before the Supreme Court of Colombia in 1982. On the day on which the final deliberations of the twenty-plus justices was to take place, the M-19—a Cuban-oriented terrorist group—seized the Justice Palace, assassinated twelve of the justices—the twelve members who were reviewing the extradition treaty—which was going to be declared legal and constitutional, and burned all the records in the Justice Palace of Colombia that pertained to drug trafficking and drug trafficking cases.

They have assassinated more than twenty other judges, and have gone so far as to assassinate one of the judges in Houston. He was sent a coffin to his home in Bogota, Colombia as a threat. The man had a heart attack, was rushed to Houston for emergency open heart surgery and, in the recovery room his oxygen tubes were cut and he died instantly.

So, this is the type of campaign that has been carried out by the Medellín Cartel. They have assassinated top military leaders, top police officials, and so on. And we saw with the Justice Palace incident, the collaboration of Soviet irregular forces—the narco-terrorists. Not only is the M-19 involved, also the FARC, the Moscow-oriented guerrillas of Colombia, who are connected to the Communist Party, who have publicly stated that they are organizing the coca growers, that they think coca growers should take advantage of the boom that is taking place in drugs right now.

There are thirty-two different guerrilla fronts and at least half of them are in coca-growing regions. They partake with the mafia of the profits. There have been disputes between the guerrillas and the mafia, but these are labor-management disputes. There are disputes, in fact, as to whose plane was shot down, because it's a very complex business arrangement that they have. That is, the planes come back from other countries loaded with arms for the guerrillas. Therefore, when the planes are loaded with drugs going to the United States or elsewhere, they are the mafia's planes. When they are loaded with armaments or other supplies necessary to the guerrillas on the return trip, they are the guerrillas' planes.

The Medellín Cartel is now an insurance cartel that is not involved in the actual physical shipment of drugs. Pablo Escobar Gaviria, for example, doesn't deal in drugs. He's an insurance man. He insures every shipment of drugs from

Fernando Quijano

South America to the United States. He gets a cut from every shipment that gets through, and makes good on whatever shipments are lost. They've also diversified, so that now they have a system where the local person can invest in a shipment, take a little piece of the action. If the shipment gets through, they get paid off a certain amount. Actually, there are celebrations every time a shipment gets through. You can tell in Medellín every time a shipment gets through, because there is a celebration with fireworks.

This is all a big publicity campaign, actually, to try to get support from the population for their activities.

And, finally, of course, we know that Fidel Castro, in private conversations, himself has stated emphatically that this is a boom of which Latin America should take advantage since it will not last long, since synthetic drugs will gradually replace cocaine in the U.S and European markets.

## The Role of the IMF

Where does the International Monetary Fund come in? Colombia is the only country in Latin America that has received actual bank loans from international banking institutions over the last two years, and they have been huge. They have received $2.26 billion of loans over the last two years. The reason Colombia receives these loans, and no other country, is because Colombia's reserves are huge compared to its debt and compared to the size of the country—they are over $6 billion, and growing.

Colombia is the only country in Latin America where the price of the black market dollar is cheaper than the official government dollar. That is, in most countries in Latin America, because of IMF devaluations, if you want to buy dollars officially, they sell for a very low price, but you can't buy them from the government, so you have to go to the black market where they sell for an astronomical price compared to the official price.

In Colombia, there is such a profusion of dollars that they are actually cheaper in the black market. That is, drug traf-

157

fickers are trying to launder as many dollars as possible, so this occurs. And everybody has pointed out, including key Colombian bankers and the European financial press, that if it were not for the drug traffic, Colombia would not be able to pay its debt. In fact, only three weeks ago, one of the heads of the local banking association, Fernando Londoño, stated that this is the first time that a country has gotten a huge loan with its conditionalities being that the well-being of the mafias be guaranteed. And that is a fact. Because if the war on drugs were to really take off, if extradition were to continue, if all the military, political, and economic measures were taken to stop the flow of drugs, then Colombia would immediately stop receiving about a billion dollars that come in through the sinister window, where pesos are exchanged for dollars with no questions asked.

This is while you have the most dramatic poverty, drop in real wages, destruction of all infrastructure, and so on in Colombia. Nevertheless, on paper, the objective monetary economy appears to be very successful, and the bankers are willing to lend, and the Soviets partake in it through their military and other involvement in the drug traffic.

## Where Is the U.S. War on Drugs?

And what's the situation in the United States with the war on drugs? Last week it was announced that there would be a 55 percent cut in the budget of the Coast Guard. The Coast Guard will have to curtail its anti-drug interdiction efforts by 50 percent. President Reagan, in the State of the Union address, also stated that now the war on drugs would go into a different phase, in which stopping consumption would be the goal, and interdiction would be deemphasized. In other words, the current war on drugs in the United States is reduced to Nancy Reagan's, "Just say no."

It is being openly said, these are messages which are very clear, that no longer will the United States invest huge amounts of money, men, lives, and so on, in trying to stop the drug traffic. That's just what has happened—imperceptibly, qui-

Fernando Quijano

etly, without anything being said over the last several months.

One should not think it fantastic that this is somehow the result of a deal that has occurred between Reagan and Gorbachov. It is well known and it's now in history books, and it's a fact, that when Kissinger carried out the opening toward China, or the famous China card, Kissinger instructed all U.S. intelligence agencies, the State Department, and all government agencies, to remove China from the list of countries that were involved in narcotics trafficking. So, from that time on, China was no longer involved in drug trafficking, officially. Why? Because Kissinger said so, since now the China card was going to be played.

This is essentially what is occurring now. Now, we're in a new process of peace—the INF talks, etc. The Soviets are no longer involved in narco-terrorism, and trafficking, and therefore, quietly and almost imperceptibly, but nevertheless very successfully, the whole effort is being scaled down.

This sends a message to Colombia, to the Colombian authorities, and to many, that they can be comfortable in being gutless, in facing down the drug mafia. That is essentially what the Colombian government has decided to do over the last period. They have decided that they're not going to extradite. There is even talk of a plebescite as to whether or not to extradite. This is the equivalent of taking a vote as to whether something is good or bad—the question of what is or is not moral is being put to a vote! This is being done right now.

I think it is imperative that we, in the Schiller Institute, the people who are attending this conference, be absolutely clear that the war on drugs is something that has to be fought, totally; that drug trafficking is completely tied to the IMF, and it is completely tied to the question of what kind of a New World Economic Order we want. The destruction of most of the West's youth, the destruction of future generations throughout Latin America is occurring, in a dramatic form.

Drugs present a real strategic threat, because when you're talking about 400,000 individuals involved in Bolivia in the coca economy; when you're talking about other hundreds of thousands in Colombia and Peru involved in drug trafficking, in one way or another—peasants involved not in trafficking

159

but in the economy of it—then you're talking about the possibility of Soviet irregular warfare forces mobilizing an army of tens of thousands along the South American-Andean spine. This is a real strategic threat in the future, and of course, it destroys any real possibility of a just New World Order.

This is what the IMF has done. This is what is being supported by, and being carried out by, Soviet irregular warfare, with U.S. government complicity.

I think it is very important that we understand that our concept of man is totally different from that. Obviously, the Soviets view human beings as something totally different than we do, because otherwise, they would not use drugs, in their efforts to expand their empire, to destroy an enemy. Obviously, the IMF and the usurers would not destroy whole populations in order to continue their debt collection, if they did not share the Soviet view of man.

Pope John Paul II said in Colombia, when he visited this last time, "slavery has been abolished throughout the world, but new and more subtle forms of slavery have emerged. Today, as in the 17th century, the lust for money has seized the hearts of many persons, turning them through the drug trade, into traffickers in the freedom of their brothers, and enslaving them with an enslavement, more fearsome at times, than that of black slavery. The slave dealers denied their victims their exercise of freedom. The drug traffickers lead their victims to the very destruction of their personalities."

This is a really fundamental question in terms of the dignity of man in the New World Economic Order. This *is* worse than slavery. It's something we must face down and destroy. Thank you.

RICARDO VERONESI, M.D.

# The Battle Against the AIDS Epidemic

Mr. Chairman, Mr. and Mrs. LaRouche, ladies and gentlemen. Let my first words be my expression of gratitude for what these organizations, *Executive Intelligence Review* and the Schiller Institute have done for us in Brazil in the last two years, when we organized international meetings on AIDS, and their invaluable cooperation in those meetings.

I am going to present my experience in this problem, as a public health officer in Brazil, as a professor of infectious diseases at the University of São Paulo, and also as president of the Brazilian Society for Infectious Disease.

AIDS is actually a real pandemic. It spread from Africa all over the world. In absolute number of cases, it is second in rank [only to malaria]. We have 3,000 reported cases, and, due to under-reporting, we estimate we actually have 15,000 cases, including full-blown AIDS, ARC and other forms.

My concern is the pattern, or model that will be followed by AIDS in the Third World. It's my feeling that the model which will be followed in Latin America and in Asia, will be the same model that is already in Africa, which is completely different from your American model. However, there are char-

Ricardo Veronesi, M.D., *is from Brazil. He is president of the Brazilian Society for Infectious Disease and a professor of infectious diseases at the University of São Paulo.*

acteristics which make the AIDS model in Africa completely different from the model that will be followed in the rest of the Third World.

First, the African model includes monkeys, which originate the virus, as a reservoir, and then heterosexuals, male and female one-to-one. This proportion in Africa is not followed in the United States or in my country. In addition, we have many more cases of children with the disease in Africa and in Brazil, than in the United States. Also, while there is an economic impact of AIDS in developed countries, you can imagine what will happen in the economies of the underdeveloped, or developing countries, when AIDS spreads all over the Third World.

AIDS is starting in Asia, with only a few cases, in India and other countries. But very soon, and this is a concern of the WHO [World Health Organization], the problem in Asia will be much worse than the problem in Africa. You have to remember that two-thirds of mankind lives in Asia.

## Preventive Medicine Too Costly?

There is a cycle of health and economics, measured by the WHO, which shows that countries that do not invest in health, and in preventive medicine, have to spend much more in treatment of disease. This involves sanitation and low salaries.

We already have endemic, chronic diseases which are out of control. In my area, in Brazil, we have schistosomiasis in millions of people, chagas disease in millions of people, malaria, tuberculosis, and the worst disease—famine and starvation—that kills half a million children from birth to five years every year in Brazil. The biggest rate of infant death in the Americas occurs in Brazil.

The governmental policies of the Third World are taken as a model in Brazil, and in the Third World, health is not considered a good investment. There is no fast return, during the governor's term of office. Sometimes to have a return on the investment in health, it takes five to fifteen years, and this does not work for politicians or demagogues. So they do not invest in preventive medicine; they prefer to invest in hospitals and treatment, which is much more expensive than preventive medicine.

The budget dedicated to health in my country is around 5 percent of the GNP. This is not enough even to control our endemic, chronic diseases. So, AIDS is not considered a priority by the minister of health of Brazil or the Brazilian government, because they don't have enough money to invest in this disease. And they don't have enough money to invest in other chronic diseases, which have already gone on in Brazil for many years.

Also, there is religious, mainly Catholic, involvement in AIDS in my country. This is due to the Catholic point of view against homosexuality and the advice to use condoms, which they oppose. So there are conflicts with the Church, when we start campaigns to control AIDS.

When I was health secretary to the city of São Paulo, the fourth biggest city in the world (it will be the second biggest at the turn of the century), the first action I took was to make mandatory blood tests in blood banks of the municipality of São Paulo. This was the first official action in an effort to control AIDS. Actually, only three days ago, there was an act of the President, making mandatory, by law, the screening against AIDS in blood banks. Only 30 percent of the blood banks actually make the tests in Brazil. But, it's not enough to make the law in developing countries; the law does not decide—it's not enough. Only long-term education and information make this action effective.

## Insects in AIDS Transmission

There are a lot of aspects of AIDS to discuss, but I'm going to bring to this audience one that is very controversial. It is the role of arthropods—flies, mosquitoes, insects—in the transmission of AIDS. The point of view of the CDC (Centers for Disease Control) here in the United States is completely against this—they do not admit this possibility.

One of the strong arguments they use is that if AIDS could

be transmitted by flies or mosquitoes, why aren't children affected equally with adults? These people don't know that transmission by arthropods is not prevalent in children, but is prevalent in adults—it is elementary in epidemiology. If you take as an example malaria: The group of age zero to five, or five to ten with malaria is very small. But the CDC does not even make a comparison with other diseases which are transmitted by arthropods, like malaria. They say, if AIDS is transmitted by mosquitoes, children should be infected in the same way where malaria is transmitted by mosquito. No one would deny that. But if you go to the group of children, you will see .5 to .7 percent are infected, just as in AIDS!

Also, they cover up a lot of data from Belle Glade, Florida, where 15 percent of the population, which works on sugar cane plantations, test positive for AIDS, and do not belong to any risk group. This is an area where the arthropods, mainly flies and mosquitoes, have a very high density.

Also, they say that in Africa, children are unaffected. Yet, the Panos Institute published a book on AIDS in the Third World, and they say in Gabon, with 5 million inhabitants, there are 6,000 children with AIDS. And still, CDC denies that there is an AIDS problem for children in Africa.

Maybe it's not a problem for children in the United States, but we heard just a week ago, that one out of sixty newborns in New York tests positive for the virus. One of every sixty newborns—this is tremendous, this is something to think about!

Maybe there are many areas in the United States, where the mosquitoes do not have a role in the transmission, but there are areas we have to consider in the United States, which behave like the Third World, such as Florida. This is like many areas in Latin America, in Brazil, in Africa, where the conditions are sufficient to accept the role of mosquitoes in the transmission of the virus. You must have a density of humans being infected to carry the virus in the blood; you must have a density of arthropods, for the possibility and probability of transmission. These conditions we have in Africa, and in a few places in the United States as well. We have it in many places in Brazil.

Ricardo Veronesi, M.D.

# Is the Truth 'Inconvenient' for the CDC?

But there is another strong argument in favor of the transmission of the AIDS virus by arthropods. There is a disease—
known for a long time, more a century—in horses, which is
called infectious equine anemia. This disease is caused by a
virus that belongs to the same subfamily of retroviruses—
lentivirus—as the human retrovirus that causes AIDS. The
same subfamily causes the AIDS disease in horses, with im-
munosuppression, and everything as in humans, and it kills
horses epidemically. You know how this retrovirus is transmitted from horse to horse? By fly bite, by *tabanedae*, the
family of flies that transmit the virus. They bite one horse and
transmit to the horse nearby.

There was a meeting organized by the U.S. Congress last
September that admitted the possibility of arthropod transmission, and they advised the CDC to follow this study, instead of being aggressively against the hypothesis of this
transmission. Many other viruses are transmitted by flies, by
mosquitoes. Why not this retrovirus? Just because it's not
convenient to the CDC?

The reason is, when it's confirmed that AIDS is transmitted
by flies, this is a problem that is of concern to the public
government, because they are responsible for the conditions
of environment that allow this population of mosquitoes or
flies to proliferate. If the problem is only homosexuals, or drug
addicts, they say, that's your problem; you're homosexual, it's
your problem. But when it's flies or mosquitoes, this is a
problem that the government must solve: Why is there such
a density of mosquitoes or flies, or other arthropods?

Another aspect of the disease is the presence of HIV-II,
which was first detected a few months ago in Portugal, in
people coming from Africa, then in six different European
countries. Then, we detected it in Brazil. It was reported in
August in the medical journal *Lancet* in England, and I have
a copy of this.

I was surprised to read in *USA Today* yesterday, about a

"new AIDS strain—New Jersey doctors have diagnosed the first AIDS patient in the Western Hemisphere apparently infected with HIV-II, a new form of AIDS." Well, I sent the CDC a copy of my research study, and I sent them a copy of the *Lancet* publication, and they just ignored it. They said for the first time yesterday, a case of HIV-II was diagnosed. Do you know why? Because the CDC doesn't like my position on arthropod-borne transmission, and so they didn't publish the data I supplied to them in their monthly publication. This is the result of politicians involved in public health, which is very bad, all over the world.

Finally, I wanted to make my prediction for AIDS in the Third World. In my opinion, it will be apocalyptic, and in fifty years, mankind could be drastically reduced, if this virus does not become naturally attenuated in mankind, or if mankind does not change its lifestyle, and government policies don't stop covering up the problem, mainly in developing countries. Thank you.

JAMES W. FRAZER

# Medical Research in a New World Economic Order: An Uninhibited Scientific Future

Mr. and Mrs. LaRouche, Mr. Chairman. I was asked to have a rather uninhibited discussion of what kind of science we will have twenty years from now. I'd like to say right off the bat that I've been in a number of planning sessions with the federal government, where the administrators would make their wishes known, and the policy makers would give their judgments, the budget officer would tell us all what we would do, and then everybody would leave. The scientists and engineers would sit there looking at each other and wonder, "How the hell are we going to clean up this mess?"

I'd like to point out to you that Mr. LaRouche, who is sitting right here, as he has sat through many, many seminars, and Helga who is sitting right there, have both contributed and have proved that there is no such thing as a separation of economic policy and scientific pursuit. You cannot have one without the other.

I am reminded of the story of Gilgamesh. You remember that Gilgamesh was sent down by the Lord to do all the heavy

James W. Frazer, Ph.D. *is a consultant and adjunct professor of pharmacology at the University of Texas Health Science Center at San Antonio.*

167

labor, and he did very well, as long as he had ethical instructions. I submit to you that the modern scientific community is very much like Gilgamesh. As long as they have ethical instruction, they will produce things for the public welfare. But I point out to you that, at the present time we spend more than 20 times as much on killing people, as we do on giving them life. This, I think, is an ethical issue, which we'll have to cure, or there won't be any 21st century science, or much of anything else.

## The Future Is Not Uninhibited

Now, having said that, the future is not uninhibited. We have a combined public-private indebtedness in the United States of something like $12 trillion. Twenty-five percent of our national budget is spent on debt service. The cost of science is borne out of free funds—funds not obligated for other fixed costs. That means that by 1996, the cost of debt service will be sufficient to shut off large science in the United States.

That doesn't mean that all sciences stop. That means that guys like me, who are used to working by themselves, and doing independent basic science, will go on working. But it means that the large applications of basic science will come to a screeching halt. You should consider that.

Scientists have learned a few things about economics. They have learned that promotion and retention at any major American university depend on getting money for the administrators to administrate. Administrators don't care as much where the money comes from, but it helps your promotion if it comes from the federal government. Federal funds, on the other hand, are mortgaged, and this year, the ratio of unsuccessful grant applications to successful grant applications will be almost ten-to-one. That means that you're going to have nine very dissatisfied people for every person who has actually got work, because those people are going to have to get out of the scientific profession.

Private industry is used to picking methods, materials, and

products out of the literature for free, and that literature is kind of interesting, in view of what's going on in Boston right now. Successively, the proceedings of the National Academy of Sciences, the American Academy for Scientific Advancement, the National Geographic Society, the American Association of Biochemists, and the American Association of Physiologists, all have been hauled into court by the federal government, called a profit-making institution, and their publications are now taxed as a regular newspaper or magazine is taxed.

Consequently, where you used to get a small remuneration for publishing a paper, or, at the most, they would get published for free, if you publish a paper in any scientific journal now, you have to make your own figures—this costs about $500; you have to pay $160 per-page charges; so the total cost of just publishing the research, not doing it, just publishing it, is between $1-2,000. The publisher then has to pay taxes on all of this, and, of course, the $1-2,000 has to come from some place. It doesn't come out of my pockets, because my pockets aren't that full.

So that means that all of that money has to come out of institutional grants. The institution gets 50 percent right off the top of the grant money to put into institutional funds. When you have a grant actually given to you for doing any kind of research, that means that you are going to have, after the cost of all the people is taken out, roughly 8 to 15 percent of the grant to get the equipment and go to work—8 to 15 percent of the total amount of the grant is what you actually have to use to do some work.

Science, obviously, cannot go on this way. There are two solutions. One of them is to get the federal government out of the way, to quit impeding the progress of people who are paying their own way; or, it will become a pay-as-you-go business. Those who use the product are going to have to pay for it. You don't have any other choice.

Students are bright people. The brighter students who get interested in science, would work in science for nothing, if they were allowed to work on a project to its completion and to its intellectual satisfaction. I have had several summer students who have worked 12-14-16-18-hour days for months at a time, no days off, to get a project done, and they did not get paid one dime. This has been going on for more than ten years.

If there is an intellectual satisfaction in the work, the cost of the work in dollars goes down. Because people will then volunteer their efforts, because they are interested in the project.

But, students are very bright people, and only the very brightest get interested in real, basic science. When they see the plight of people who are administratively hampered from obtaining that intellectual satisfaction, when they see that remuneration depends on the same kind of activity that a used car salesman uses, they can also see that it isn't worth the trouble. Consequently, U.S. students are avoiding science in droves. Our better students are going off into business administration or something where they can put in a nine-to-five day and get out of the office. For this I really can't blame them very much.

## We Are Losing Science Students

Either we are going to have to learn how to allow intellectual satisfaction from a good piece of scientific work, or we have lost our scientific future, because a student simply will not put up with it.

The other side of that coin can be seen in all of our graduate schools. I speak from my own experience from the University of Washington at Seattle, the University of Colorado at Boulder, the University of San Diego, the University of Texas at Houston, Austin, and San Antonio, the University of Florida, the State University of New York, and from Purdue University. All of these places have a heavy preponderance, two-thirds of the graduate students, of foreign students. They have competitive examinations in engineering and physics departments, in particular. The reason they have those competitive examinations for entry, is because they had such bad experience with the caliber of U.S. undergraduates in the early 1970s.

The persons taking those competitive examinations are now almost uniformly beaten out by graduate students coming from China, India, England, very few from France, and none from Germany—Germany sends their post-doctorals here. That means, of course, that we are very good in doing what we do in graduate schools, but it means that very few of the products stay in the United States. We are actually training our scientific competition.

As a matter of fact, it's even more humorous than that, because some of the best students I've had were at the University of Colorado, who were sent here from Russia. Let me remind you that the University of Colorado is very close to some of our nuclear storage depots, so they're learning a lot more than engineering when they come here.

# The Future of Science

What kinds of science can we expect to see in the future? Let's assume that Mr. LaRouche becomes President, or at least we get somebody a lot more sensible than anybody else I've seen recently. What kinds of projects will we do?

I'm going to do this in the way I want to do it, because it's my belief, and I think, demonstrated experience, that basic science precedes any application. Our present practice of saying, "We are going to work on this, that, or the other project," you fund the project, and then three percent of what you funded in the project winds up in basic sciences, is a despicable scheme, it's dishonest.

What you need to know for the major health problems, is what's the difference between a cancer cell and a normal cell, and how do you make it like a normal cell. Our present technique is to screen a whole bunch of drugs and try to kill both cells, but leave a few of the other ones alive, so the person can stay here. We have now poured $30 billion into that kind of philosophy, and I would think that, after a few of those dollars, we would learn that that is not the right way to go about this business.

On the basic scene, Dr. Rinaldo Dulbecco at Progresso Cultura, a meeting held in Milan last summer, said that he thought that one of the most important things to do, was to get the complete sequence of the human DNA genome. But, he also said, present methodology is not up to the job. The methodologies that look best are far on the horizon. These are particular kinds of spectroscopy, such as one that Jim Gregg has out at Los Alamos. This is a special form of optical biophysics, using a laser back-scatter measurement.

Another kind of work that would be profitable would be that of Fritz Popp, who is measuring the optical emissions from a molecule going through different transition states. He has actually measured emissions from unfolding DNA. One of the things about Dr. Popp's work which has us all a little nervous, for a number of reasons, is that that light comes out coherent, with very long, coherent path-lengths. It looks as though DNA and several other molecules in the cell are acting like excimer lasers. I'll have a little bit more to say about that in a minute.

If the complete sequence of the human genome were to do us any good, we would have to understand what it did, how it was controlled, why all the cells do not come out the same, but adapt to their different environments and perform different functions.

## Central Theory of Biology

One of the things which has been missing for many, many years is a real central theory of biology. It is now appearing that this central theory is going to revolve around the complementarity of different systems, where you have a complementarity fit at the cell surface, another complementarity fit to the RNA going back to the nucleus, another double complementarity within the nucleus and then a series of complementarity fits going from the nucleus to the cell cytoplasm. You have these levels of complementarity.

# James W. Frazer

The exact structure, geometry, and rates of mobility of those complementary systems will probably occupy an awful lot of work for at least the next ten, and probably the next fifteen years.

Another kind of activity, completely shifting our attention for a moment, is the recent announcement of superconductivity in ceramics, the so-called room temperature conductors. These superconductors are at once a very practical regime, everybody talks about levitating trains and this sort of thing— well, let me tell you, that's some time off. But the thing to keep in mind is that the appearance of the room temperature superconductor was a victory for the experimentalists. There was no extant physical theory of condensed state that could have predicted that these would be possible or occur.

Furthermore, their appearance says that you can have magnetic lines of force induced by a current in a superconductor, but the process of levitation, called the Meissner effect, says that the magnetic fields of force can't enter the superconductor. This is a direct violation of Gauss's theorem, which was the basis of the First Law of Maxwell for electromagnetic propagation, and shows that we have a fundamental misunderstanding of the nature of electromagnetics. This is one of the reasons this has excited so much physical interest; it is a direct contradiction of some of the most fundamental of our physical understandings.

Now, let's go back to biology for a minute. We have found that the effect called the nuclear-Overhauser effect, which means that nuclei can couple together with a magnetic field, directly through space, not dependent on chemical binding, is very prevalent in biological systems. It is so prevalent, that if one does a nuclear magnetic resonance spectrum of biological tissues, you usually get just a big swoop; because as soon as you excite one part of this molecule, it couples to almost every other part of the cell. You can do this to hydrogen protons, you can do it with P-31, you can do it with calcium. You can have a good time, except it gets to be four o'clock in the morning and it's time to go home.

In conjunction with this apparent low-order superconductivity in biological systems, comes a set of very curious membrane receptors. These are on the external surface of the cell, but they're manufactured, of course, right next to the nucleus. They have a protein base, which makes the biochemists happy. But they have an immunologically-coated sugar-coat, put on them before they're put on the cell surface, and that is what you recognize with the immune system.

We know that the AIDS viruses recognize something called the t- or thy-antigen on the surface of the cell, and that antigen has got four different protein strings hooked together. Each string has a different sequence of carbohydrates attached to it. Would you believe that no less than eleven different chromosomes code for different parts of those strings. Parts of those strings are a stable configuration. Parts of those strings actually react to the presentation of an antigen. And you have an almost infinite code-shuffling going on in the nucleus, so that you could react specifically with external antigens.

## The AIDS Problem

That's the kind of a problem that the AIDS virus really poses for us. How does this virus control the nuclear reshuffling which produces these antibodies? That's the $64 question. I think, maybe, we may have two sets of drugs that can stop the AIDS virus from preventing the reshuffling. It doesn't cure the disease, but let's you live a little while.

As far as fundamental understanding of this system is concerned, we might get there in twenty years, and we might not. But, we're never going to find out unless we do the work.

The occurrence of these receptor cells is not, of course, limited to the immune system. In our brains, we have receptors, little molecular receptors, all over the surface of every one of our neuro-cells, and also, several of the so-called nurse cells that support nerve cells. The actual activity of the brain is caused by neuro-transmitters, chemical substances being released from one axon, and this takes about a micro-second to get to the following point, and what happens then depends on what kind of neuro-transmitter lands next to it. We have over 100 different molecules that can act as neuro-transmitters.

If you do 100 factorial, that's the total number of possibilities for coding. If you take that factorial, times the number of cells, times the area of each cell that can be occupied, you come up with something on the order of 10 to the 55th power events per second, which your brain is now performing. Our chance of attacking that type of problem in completeness is not possible until we have a computer a lot bigger than any that I've seen recently.

Howsomever, we do have the major pathways through neuro-transmitters laid out, we know where they synthesize, we know how to make NMR pictures of it. There is a new technique which is called flash MR, which allows us to get the chemical identifications. One can expect that we'll have the definitive locus of thought control within the brain, within about the next fifteen years.

## Achieve Genetic Control, or Die

We have then two systems wherein we can achieve genetic control: one of them from the surface reaction, like the AIDS virus; the other one within the brain itself, like controlling fuel distributions.

The society which learns how to control these processes first, will control the world. The society which does not invest and learn how to control these processes and develop the ethical principles to control the knowledge of these processes, is going to die. It's that pure and simple. Either you find out the process, learn how to control it and survive, or your society is going to die, just as certainly as we're sitting here. We don't have any choice.

I visited France last December, and was privileged to talk to Mr. de Loch, who is the assistant research director for the French Atomic Energy Commission. You have all heard things about fusion research for about the last thirty years. You know we've been trying and trying and trying, and that the thing to do is to cause a small pellet to explode, but not explode very far out, so that it gets a lot of energy, but you have to keep it packed, so they can then fuse with each other.

The French, by using a very high-powered gyrotron, inside of a tokamak and a laser pulse on a small hydrogen bead, are right on the dividing line for net energy production from controlled fusion. They expect to be at commercial levels within the next four or five months, and then comes the business of building controlled fusion reactors.

The two key elements that were used to produce this energy were a gyrotron, a Russian invention for the generator for the fields, and the tokamak, the most improved tokamak.

Now that we have superconductors coming along, the fields inside the tokamak will go up, and actual fusion ignition and power production is a possibility within four or five years.

A friend of mine, named David Cohoon, who is, in my estimation, the world's most complete mathematician, has recently examined the field of electromagnetics, mathematically, using some rather advanced Halburg spaces that I don't even pretend to understand myself. Dave comes up with a set of eleven different differential equations, which are necessary to operate simultaneously, to describe electromagnetic propagation. If one adjusts the parameters in his equation to match the conductive properties of the superconductor, you find that you can have magnetic field traverse of a superconductor at six times the speed of light.

Ladies and gentlemen, welcome to the way to get to the stars.

DONALD ERET

# How International Pricing Systems Have Destroyed Agriculture

Mr. and Mrs. LaRouche, Dr. Wills, ladies and gentlemen: The topic I will discuss this afternoon, the international pricing systems of agricultural commodities, is based on the premise that the prices of agricultural commodities worldwide find their basis in the United States, primarily through two institutions—one being the federal farm programs, in establishing the commodity loan rates, thereby establishing the floor prices of commodities, and, traditionally, setting the cash prices. Over the years, however, these loan rates have been on a decreasing scale, continually.

The other basis of pricing is through the commodity exchange boards, where the practice of commodity speculation continually has a downward pressure on the prices, has forced them down to the established commodity loan rates, and now, in more recent years, has pushed those prices on down below that price.

Donald Eret *was a Democratic member of the Nebraska State Legislature, from 1983 to 1986. He is a mechanical engineer and a farmer.*

## Deteriorating Farm Situation

Along with a skyrocketing national debt, increasing foreign trade deficits, national industrial cutbacks, and reports of domestic and worldwide hunger, an unsettling era continues to develop in U.S. agriculture. Reports of agricultural economic problems in other countries around the world suggest that some common cause is to blame, that this is no coincidence.

Grain and livestock commodity prices have not kept pace with the increasing costs of production. Land values spiraled and have now dropped to less than half their peak values in 1981, bankrupting thousands of farmers, for whom it was their time to make a commitment to farming. Bankrupted or severely strained also are financial institutions, who had backed them. The land is not abandoned; it is taken over by larger and many times more leveraged operations. Farmsteads where people lived and husbanded livestock and which were the making of a neighborhood and a community, are steadily being abandoned, either to decay, or settings are razed to be removed from property tax rolls.

The effects are most notable along the main streets of the hundreds of rural towns, with their many empty business buildings and stores. Farm equipment dealerships are rapidly disappearing as sales have dried up. The equipment manufacturers go out of business or are merged and discontinue proven lines of machinery manufacture. School districts continue to be consolidated. In Nebraska seventy-five of the ninety-three counties are experiencing a continuing population decline.

In contrast to the World War II period and years thereafter, when every farm was a fully-programmed operation of beef, dairy, hog, and poultry production, the countryside is now a haunted scene of selective specialization. Grain production is highly mechanized, however, now with aging equipment, but is carried out under a government program approaching 35 percent shutdown of an already restricted authorized acreage base, established from past cropping history.

For those commodities that are harvested and produced,

178

the storage and reserve stores are being classified as surpluses. Government loan rates that had historically been relied upon to establish an adequate market price, have been replaced with target prices per unit produced, which are now the basis for receiving subsidy payments from the government. The price-setting loan rates have been reduced to 50 percent of parity or less, on the pretense of underselling foreign producers. This causes commodity prices in the other countries to be reduced, and then requires their governments to budget large subsidy payments to their farmers. Cheap commodity prices also allow processors and exporters to effect better mark-up profits.

The current presidential campaigns are bringing out varied agricultural goals. One Democratic candidate promises higher commodity prices, with mandatory production controls. Republican candidates, who follow the current administration line, are going to a free market system, along with decoupling of subsidies for their ultimate suspension. On the other hand, Lyndon LaRouche proposes full production for the world need, at parity price under a mobilization plan. My following section on commodity speculation will explain why a free market system is not a supply-and-demand system of marketing.

There is some misunderstanding of what farmers mean in recommending mandatory production controls. It does not necessarily mean restriction of production to create a shortage in order to raise prices, although that would work to a certain extent. Reference is most usually made to previous periods when government commodity loan rates were set at 90 percent of parity, and the government managed production levels to meet required needs, and pro-rated these projected production levels, plus an adequate reserve, on the basis of an average county yield to each producer as an assigned acreage to plant.

A new administration employing mandatory production control, could not very well exceed the no-plant, set-aside quota that is now being implemented. That, on top of the Conservation Reserve Program goal to take forty-five million acres permanently out of production. On the other hand, caution must be taken not to repeat former USDA Secretary Butz's call of 1973 to plant fence row to fence row, including marginal land, which precipitated the land price boom that has since collapsed with devastating results.

The large farm program subsidy payments that are being paid for not planting a specified number of acres, should instead be diverted to strengthen the successful Public Law 480 Foreign Aid Food Assistance Program and the domestic food help programs of processed commodities distributed by the USDA, in cooperation with state and local governments. These programs increase the need for produced commodities, employ people to process them, add the required diet fortification needs, and are processed and maintained to government dietary specification in sealed cans or cartons. This is a need that definitely needs to be administered.

## Commodity Speculation

The 1987 approval of quadrupling the speculative position limits per trader from 3 million to 12 million bushels of corn, wheat, and soybean futures contracts by the Commodity Futures Trading Commission (CFTC), at the request of the Chicago Board of Trade, was a move in the opposite direction than many producers and their farm organizations wanted the commission to take.

Since the late 1970s, producers have been calling for adhering to the legislated provisions of the Commodity Exchange Act of 1936, which was referred to in the Congressional Record as an act to limit or abolish speculative short-selling of futures contracts.

Each commodity futures contract has a seller and a buyer. By law, the commodity brokers keep account records, of whether the contracts are commercial or speculative. When the accounts of open interest in the futures contract exceed specified commodity amounts, they are reported to federal regulators by the brokers on a monthly basis. Speculators are to abide by a limit of contracts, that they can hold for each commodity, and this is referred to as a speculative position limit. There are basically four positions that can be held in a commodity contract: a commercial buy or sell, more commonly referred to as the long or short position; the speculative long or short position. The prices per bushel, per hundred weight of commodity that these contracts are traded at, on a U.S. commodity exchange set the basis for the cash sale prices for the commodities worldwide.

What should concern everyone, and does concern the producers who understand the dysfunction of futures trading, is the accumulative volume of speculative short sales that are initiated in one year, relative to the annual production volume of the commodity. Each such contract, displaces a like amount of actual commodity from having a producer represent its sale under an equitable system of supply-and-demand marketing. Each such contract is made by the speculator, in hope of making a profit on a reduced price, and a large volume of these types of contracts does force the price down.

The real damage occurs when the annual cumulative speculative short sales of contracts of a commodity approach or exceed the annual volume of production of a commodity, and thus leave the producers without any effective marketing representation of any of the commodities they produce. Common sense, intuition, and logic tell one that this just is not right. It constitutes an inequitable marketing condition.

In a recent one-year period, the estimated speculative sale of soybeans was 2.7 billion bushels, which exceeds the annual U.S. production of 2 billion bushels. Wheat, at 1.3 billion bushels of speculative short sales at the Chicago Board of Trade, represented the major portion of an annual U.S. wheat crop.

The Commodity Exchange Act allows the speculative position limits to be established at separate levels for long and short speculation. However, to date this has never been done.

The way that bear raid speculators drove down the stock prices on Black Monday, Oct. 19, 1987, by short-selling the stock index futures, is typical of the way that agricultural commodity prices have been battered down by speculative short-sellers for the past decade.

After the large U.S. grain sales to Russia in the early 1970s, Congress removed commodity futures trading regulation from the USDA, on the premise of relieving the department of a conflict of interest in managing production and in regulating marketing, and thus established the Commodity Futures Trading Commission [CFTC]. In short order, the new commission, headed by presidential appointees either from or endorsed by the futures trading industry, had reinstated options trading that had been outlawed in 1936. It has continually relaxed trading rules, and has brought in the securities index futures contracts. The *Wall Street Journal* of Nov. 12, 1987, reported that following the filing of a bill by Rep. Jim Leach of Iowa, aimed at reducing speculation in financial futures, the Illinois congressional delegation vowed to block the bill. Senator Alan Dixon of Illinois, one of the leading Board of Trade political action committee recipients, stated that the Chicago Mercantile Exchange and Board of Trade are "big business in my state."

The most direct correction and solution for an equitable pricing system for agricultural commodities, lies in insisting that Congress place the regulatory responsibilities now vested in the CFTC under the Securities and Exchange Commission. Under the SEC, speculative sales are allowed only after an uptick of a trade transaction of the subject stock, in order to protect the value of the traded securities. This is the guiding regulatory rule under the SEC, which is lacking under the CFTC and will not likely be adopted by that commission.

However, aside from these considerations, there are many people who question the necessity of trading commodities at all, if our government would set an equitable price on our agricultural commodities.

In finishing up, I'd like to discuss briefly the resultant farm debt crisis, which has resulted from the lack of an equitable price for agricultural commodities.

## Farm Debt Crisis

The monumental indication of a national farm debt crisis, was portrayed by the necessity of having to enact the Agricultural Credit Act of 1987, in order to bail out the Farm Credit Banks, which are a federally assisted farm member credit cooperative. An amount of $4 billion was appropriated to shore up this seventy-year-old system, which sustained massive losses after having accumulated reserves in prior years. This same legislation also included reform measures for the Farmers Home Administration [FmHA], a federal source of lending of the last resort to assist in getting farmers established in farming. Also, debt mediation is required in both the Farm Credit and FmHA systems, under the new law.

In 1986, Congress established Chapter 12 bankruptcy, to administratively handle the monumental losses in secured loan collateral on farms and ranches due to the massive drop of land values. Since this provision became law on Nov. 26, 1986, more than 4,800 farmers have filed. Nebraska, a state that has not yet enacted farm debt mediation on commercial loans, due to resistance of the state's bank associations, leads in the number of Chapter 12 filings. A study made in Iowa reportedly indicates that 82 percent of the filers proposed to write down creditors' claims on farmland to fair market value, which has declined nearly 60 percent in Iowa since 1981.

The losses to creditors are not intended to be more than they would have been through foreclosure. Rather, it is intended to reestablish a manageable secured loan at current market value, with the written-down portion to be reclassified as unsecured. On this, the farmer will pay back as much as he can in the next four years, and what is left will then be cancelled. The average write-down of debt on the first thirty cases settled in Iowa since November 1986 was $227,500, or 47 percent of the amount owed. It is estimated that for every Chapter 12 bankruptcy, five farm loans are being renegotiated willingly by the creditors, to decrease interest rates, in order to keep as much of the loan as possible, listed as being secured. Despite these credit developments, there are estimates that over half a million farmers have been forced from the land in the United States in the 1980s.

In my conclusion, I would state that it is not beyond reality that, had Congress properly exercised its responsibility of legislative oversight over regulation of commodity speculation, as enacted in the Commodity Exchange Act, to protect commodity values, that it would not have had to appropriate $25 billion annually from farm program budgets, Chapter 12 Bank-ruptcy legislation, or the Agricultural Credit Act of 1987. Thank you.

# FRED W. HUENEFELD, JR.

# Moneygate: A Political Crime Bigger Than Watergate

The final four pages of this report are an account of the year-by-year decline in the national income share of the "private enterprise sectors" of our economy, i.e.: 1) the income of farm proprietors, 2) total income of unincorporated business and professionals, 3) the rental income of persons, 4) total corporate profits before taxes, and the corresponding rise in the primary costs of doing business, i.e., 5) wages and salaries, and 6) the interest income of individuals, as these data are provided by the Department of Commerce, in Economic Indicators and the Federal Reserve Bank's publications. . . .

Because the period from 1943 through 1952, for a combination of reasons including legislative provisions to support the war effort, provided 100 percent of parity farm prices on the average for the entire 10 years, all other sectors of the economy were in relative balance and were relatively prosperous. It was a unique period, unparalleled since 1910-1914. It was a period in which there was equity of income and equity of trade in the domestic economy, which was sustained entirely by "earned income" with an absolute minimum of excessive

Fred W. Huenefeld *of Louisiana, is former president and currently executive board member of National Organization of Raw Materials. His paper was presented to the conference for publication. This is an abridged version.*

debt injection. For this reason, we use this period as a base from which to make consistent calculations of national income share changes after 1952.

We then take each year's total national income and multiply it by the same percentage ratios that prevailed during the 1943-1952 parity period, to get a reading on what each sector would have earned, more or less, compared to what it did earn in a given year, had we been fortunate enough to maintain subsequent yearly division of national income in direct ratio to the division that prevailed during the 1943-1952 parity period. In a word, if we had been able to "stabilize" the same relative purchasing power of farm income for the subsequent thirty-five years to 1987. It is presumed that a constant 100 percent of parity on all "raw materials" would have provided that stability for the longer period, to 1987, as it had done from 1943 to 1952.

All of the conclusions drawn from this information are after-the-fact, or hindsight so to speak, because it has all happened. It is a matter of permanent record and it cannot be legitimately disputed. What we are talking about now are the adverse effects of this long-term trend, which threaten to disrupt the economic and social life of the United States for the next several years. The difference of opinion revolves around "exactly why did it happen?"

Perhaps the most important observation that can be made from this amazing array of classical statistical data is that it had been portraying convincing long-term economic trends as early as 1960, which foretold the present economic dilemma even at that early date, had we had the wisdom to react to it. . . .

Doesn't this tell us something? Doesn't it say that never before in history has our national economy performed so badly and that there has to be a logical reason? Doesn't it say that there is no denying the absolute necessity of 100 percent of parity raw material prices, in view of the fact that we only owed $485.7 billion in 1952, 175 years after the signing of the Constitution of the United States. Then, after we destroyed 90 percent of parity rigid price supports on farm production in 1953, we have had to add an additional $10,501.1 billion

Fred W. Huenefeld, Jr.

in just thirty-five years. How long must we remain married to this *perverted political economic system* before we ask to be separated from it, and how long do we have to wait before our academic economists and their universities are held accountable for it?. . .

The United States should produce for the domestic economy exclusively at American 100 percent of parity prices, even if it requires a 50 percent cut in production of farm commodities. We would be no worse off than we are today, and the United States and the world would both benefit from less cutthroat competition, and higher world raw material prices. Needless to say, perhaps, but such a move would increase consumption of food and other products produced from farm raw materials and the increased income would require both increased domestic production, but also increased imports from foreign countries—to say nothing of potential foreign exports. Underdeveloped nations would benefit directly and fully, thus (in theory at least) increasing internal demand in those countries as well as enhancing exports, with the United States out of the picture as an exporter of cheap, price depressing, subsidized farm commodities.

. . .One hundred percent of parity raw material prices are "the stone which has been rejected by the builders, but which proved to be the cornerstone" [Acts 4:11-12] of any civilized economy in the world. We have the evidence to prove it. Let's go for it.

PAULA WIEHMER

# A Call to Action

A story translated from Arabic recently came into my hands. "One evening," so the author related, "I went for a little walk. There I encountered a hollow-cheeked, hungry girl, her clothing flimsy and tattered, shivering from cold. I would have gladly given her something, but I had nothing with me, not a farthing, nor anything to eat; I had everything at home. So I quarreled with God: 'Thou surely seest she is in need; but why dost Thou do nothing about it?'

"For a while God was silent—but that night He came to my bed and said, 'I have done something about it. I created thee.' "

The story does not let you off the hook; it hits at precisely the point which we always seek to avoid, and which, out of laziness, we always want to cheat our way around. We are not passive spectators, but are all drawn into God's task of creation; we are called upon to set our talents and capabilities—be they ever so modest—toward the development of the world.

Yet why do we draw back? Why don't we act, in view of the threat of hunger, epidemics, drugs, and injustice which cries to Heaven for relief?

Paula Wiehmer *is a corporate manager in the Federal Republic of Germany. This paper was written for the January 30-31 Andover, Massachusetts conference on the New World Economic Order, but not presented as a speech.*

# Paula Wiehmer

We can find the reason for this failure to act, for this passivity, in Thomas Aquinas. "Acedia," he writes, "is that indolent sadness of the heart, that does not want to assume that greatness to which God has called Mankind; this indolence raises its paralyzing face, wherever a person seeks to shake off the duty-bound, noble essence of his real worth as a person, and even the nobility of his filiation to God, thereby denying his true self."

Over three-quarters of humanity looks toward us, with eyes of hunger. At home, in Europe and America, we have everything.

Food is stockpiled, no matter what the cost.

Food is destroyed, no matter what the cost.

Food is used as industrial goods, no matter what the cost.

But to give food to the hungry, that's too expensive.

And importing food from the starving nations is called "development assistance."

Isn't there anyone to stand up and put a halt to this insanity?

The words of the speaker from Brazil are still ringing within us. Children, whose mission is to shape the future, are instead born only to die. Drugs, which are destroying countless young people, are being cultivated in order to amortize debt—debt which was determined by arbitrary economic and financial dictates, and whose collection is therefore immoral.

At home—in Europe, in the United States—flourishing regional economies are dying, because of alleged and actual overcapacity in steel and in nearly all productive sectors of the economy. But instead of using this surplus to help the developing sector, we bear the costs of interring our own regions' economies.

Yet, if our life is dear to us, we will finally brush aside all egotistic motives. The destabilization of the Third World demonstrates the devastating consequences of not doing so—not only for this nation, but for all industrialized nations.

If mankind is to survive, we must act today; tomorrow is too late. The demand for a just world economic order comes not out of compassion, but out of prudence. Prudence, as virtue, was already part of pre-Christian thought.

The opposite of true virtue, as we read in Thomas Aquinas,

is the false virtue of tactics. It is amazingly pertinent for our present-day situation, when he further says, that all these false virtues and tricks arise from greed, and from that upon which it is premised.

Let us consider, however, that it is prudence which spawns justice. Plato speaks of something which we take in from afar, but which is fundamentally so simple, that each individual can be given what is due to him. So, when, among our various peoples, nations, races, origins, and social positions, will each be given what is due to him?

If we are to make justice into reality, it is not enough to have merely good will or intentions; there are things we must do. It is necessary for us to act, in the face of worldwide poverty, the threat of epidemics, and the danger of war. It is necessary to act, because of the challenge posed by the personification of evil in our time.

Let us venture into battle against evil's resistance, with bravery—a bravery which does not preclude sacrifices, and which, in Thomas Aquinas's view, without an orientation to the personal God, is condemned to failure.

Let us, through the right measure, ward off licentiousness in nearly all fields of life today, and let us defeat the devil's might with the strength and power of our love.

NANCY MULLAN

# Malthusianism as Economic Policy

Malthusianism is a tenet of Atheistic Internationalism, and has been enthusiastically embraced for most of this century. The 1969 *Report of the Commission on Population Growth and the American Future*, and the 1980 *Global 2000 Report to the President* never waver from the Malthusianism they spout, and helped set the stage in this country for population manipulation and decline.

In a struggle which began in the 1920s with the promulgation of artificial contraception, abortion was finally legalized in the United States in 1973.

Worldwide, it takes a toll of 53 million babies per year, including 1.5 million American babies, about one-third of those being in the third trimester of pregnancy. Early anti-abortionists often cited the danger of killing as birth control leading to killing as death control, but they were largely ignored by the populace, who were busy having their consciences further dulled by TV, pornography and rock music, while mainline churches looked in another direction.

## From Abortion to Euthanasia

Now we find ourselves on the eve of active voluntary euthanasia, with the AIDS epidemic vastly increasing the num-

Nancy Mullan, M.D., *is former President of the California Pro-Life Medical Council.*

ber of terminally ill and the economic crisis significantly increasing the impulse to have them dead. The AIDS epidemic has also been the occasion for the further promulgation of totally slanted, one-sided liberal sex education in the schools, and the further institution of condoms in minority populations.

Abortion very rapidly led to infant euthanasia; a 1973 *New England Journal of Medicine [NEJM]* issue carried the first report of forty-two children who were permitted to die at the Yale newborn nursery, because vital treatment was deliberately withheld.

## Physicians Accept Euthansia

Whatever the problem, the liberal social agenda turns out to be the solution and gets instituted. War, disease and famine are also seen as legitimate methods of reducing population, so AIDS is actually welcomed by some.

Just as it is very difficult to argue and hold out for a legitimate moral stance in the face of a totally impossible, unplanned, unwanted pregnancy, it has become increasingly uphill to support living until one dies naturally, especially if one is disabled, elderly, or has a terminal condition.

Most recently, the Jan. 7, 1988 issue of the *NEJM* contained a special, unanimous report from the Stanford University Medical Center Committee on Ethics, the tenor of which can be ascertained from the following excerpt:

". . .Physicians need to clarify the purpose of placing intravenous lines . . . It is not acceptable to begin intravenous therapy for 'hydration and nutrition' (food and water). Once an intravenous line is in place, it becomes harder to refrain from treating the infections and chemical imbalances that might provide a humane release. The same reasoning applies to laboratory tests or assessing vital signs. Finally, similar cautions apply to the placement of feeding tubes, especially in patients in chronic vegetative states."

The report also notes that once any treatment is begun, it is much more difficult to discontinue than if it had never been started.

In California, Americans Against Human Suffering, the political arm of the Hemlock Society, have qualified to place the Humane and Dignified Death Act Initiative on the November 1988 ballot. Bearing the attorney general's soft-spoken title, "Terminal Illness. Patient's Directions Regarding Medical Procedures," the measure would allow terminal patients to direct their physicians to kill them by using "any medical procedure that will terminate life . . . swiftly, painlessly and humanely."

The initiative campaign has been filled with deceptive rhetoric and misrepresentation, such as a fundraising letter which claims that patients are ". . . forced to suffer," and that ". . . even when a doctor knows that hope is gone . . . he must legally keep his patient alive on life support machines, regardless of what the patient may want."

While the public is much more sensitive to the problems of euthanasia than they are to abortion, there is grave concern that the measure will pass.

## Nobility or Tyranny?

Finally, the interfaces of birth and death have always been the occasion for moments of nobility or tyranny of men. In a century lacking almost all integrity, one could hardly hope that they would have been better handled. But this century's own technology confronts us with situations which would have challenged Solomon, especially in the areas of criteria for death and organ harvesting. Especially now, men must be pure-hearted.

The Atheistic Internationalist has world subjugation to his particular despotism as his aim, and he is not doing badly in accomplishing it. Most of America does not realize that it is currently actually enslaved by the debt money system generating big government's need to tax in order to repay Internationalism's banking elite.

Americans own less and less, opine more and more, and are fiendishly addicted to expensive habits, like using credit and buying insurance as a form of protection. They have not any notion why cash and privacy are important, think that Gramm-Rudman is a terrific idea, and have been totally easy to subjugate.

One wonders how far the deterioration will progress, and sincerely hopes that the current economic crisis will provide the pivot toward a just economic and social order.

FRED WILLS

# Closing Remarks to the Andover Conference

The British had, in the last century, a pseudo-intellectual
whose name was Matthew Arnold, an alleged writer of poetry,
who belonged to the Pre-Raphaelite Society. Arnold was one
of those responsible for this myth we call the Middle East—
that bit of geographical nonsense, according to which Morocco
is the "Middle East" although it is more westerly than Lisbon
and London. The reason I mention this is that one of the
doctrines the Pre-Raphaelites had was about Arabs. They would
quote a saying at Oxford, and at London, and elsewhere: "We
shall fold our tents like the Arabs, and silently steal away." I
always resented that.

I know their spies are here, and I know they will hear, and
I want them to know from me that this conference will not
fold its tents, like any Arabs, and will not steal away. Let them
take that back to Buckingham Palace, and to Windsor Palace,
and to Balmoral and to Barbados, where Margaret has her
own palace.

I want to thank all of you who attended this conference for
your interest, patience, and your concern. I want to thank all
the contributors who made speeches and asked questions, for
their many suggestions, and their many profound, deep, and
clear contributions. I want to thank all those who sent messages
to the conference, and all those who indicated their general
support for our aims and for our objectives.

I am no democrat—small "d," or big "D"—I am not. So, I will infer the decision that you intend that we should make a compendious record of these speeches and recommendations here. That we'll hold press conferences to foster dissemination of what we've done, and the recommendations we intend to pursue. Because what we've done, in my view, is to seek to mobilize ideas and send them into battle.

I also infer, that you want to give New England, in general, and Andover, Massachusetts in general, the opportunity to redress the immorality of what occurred from 1944 to 1946 at Bretton Woods, in New Hampshire.

We have examined the need for new and bold mechanisms of credit, in order to buttress the production of goods and services, if mankind is to survive in the present, impending, imminent disaster. We have looked at the need for technological emphasis and rational choice, in those projects that will promote generalized growth and generalized development. We have looked at the need to consign to the scrap heap, the invalid assumptions and the discredited institutions of that old Bretton Woods system, the World Bank, the International Monetary Fund, and the International Development Association.

The IDA had a curious arithmetical category—the neediest—and there were some arithmetical limits to poverty. Going to the IDA for help, we would have to forge our statistics, and people do forge statistics, to make sure that we fell into that category, to get long-term loans at small interest. So, as a Guyanese government official, I would have to come up and say, "We are the neediest," even though we were earning quite a few dollars per annum per capita. Tanzania had a problem. Their $80 per annum per capita, was just above this "neediest," and poor Nyerere saw starvation, but didn't have enough people present locally to do any sort of arithmetical calisthenics, to put him in the category of the poorest, like Bangladesh, the Sahel, and places like that.

We have focused here on the need for cultural optimism, and for revived insistence on the dignity of man, in a new international economic order. I feel, as your chairman, that we shall win, because our cause is just, and frankly, because there is, in the present circumstances, no alternative to victory.

Our enemies have found themselves impaled on the horns of a cruel dilemma: They hate us, but they know they have to come to us. That's the thing I like. They want Lyndon LaRouche's policies. They think they don't want Lyndon LaRouche. They have this curious acrobatic kind of epistemology, that can separate a man from his policy.

We are blessed with myriad strengths. We focus so much on what the enemy says about us, that we forget what we are ourselves. The chief of our strengths is the creative leadership, and the fertile intellects of Lyndon and Helga LaRouche. We have, in the Schiller Institute, the formal embodiment of the soul of the human genius. We must show ourselves worthy of such a heritage.

It has been my pleasant duty to preside over this conference, and now, to bring it to a close. But this is not a close. This is a phase change—to taking action outside. I have been sustained by your interest, encouraged by your enthusiasm. With all that lunacy prevailing outside—Dukakis and Simon, Reagan and Bush, and what have you—this room and this conference have been an oasis of sanity. I wish, on your behalf, to tell Lyn and Helga, that we intend to be worthy of their leadership. And to tell our enemies, whom we know and, as I said, are here (and they always monitor what we do; I've known them for 60 years, and believe you me, they don't give up)—that we shall never fail, we shall never falter, we shall always open new flanks, and we shall always strike mighty blows over and over and over again, until justice returns, as the imperishable axis of our human existence.

When I was leaving the university, I was asked to go and see a gentleman, Sir Jock Campbell, who is a big guy in the East India Company. The East India Company ran Boston; we have several jokes about that. The East India Company was founded in 1600, they killed Bruno in 1600, and the first settlers set sail for America a few years after that. I asked this guy, what policies he felt we should pursue. He told me, "the doctrine of regrettable necessity." I'd never heard this. But I knew the British capacity to use words, and the idea that you can hide reality, behind a facade of polysyllables.

I never understood it then, and it suddenly occurred to me when I was a minister of government. The prime minister of Guyana was always saying, "regrettably, we have to do this. We have to ban the importation of milk. Regrettably, we have to do that." And then, I understood. That is what we are trying to stop.

I feel very strong and very invigorated by what has happened today. Every moment I have spent in the presence of Lyndon LaRouche, I am quite proud to say, has enriched my existence. I don't know about you, but that's what it does for me.

This has been fine. But we have to do more. I threaten my enemies to live for another twenty years. And I intend to die before Lyndon LaRouche. It's been a good conference. We shall pursue, exploit, do what we have to do, and, have no doubt about it, we shall win! And they know it.

Thank you.

# Appendices

# Appendix A

# The Outline of NAWAPA
## Lyndon H. LaRouche, Jr.

The North American Water and Power Alliance—NA-WAPA—is the most comprehensive of a series of plans developed during the 1950s and 1960s to capture and redistribute fresh water in Alaska and Canada. NAWAPA would deliver large quantities of water to water-poor areas of Canada, the lower forty-eight states of the United States of America, and Mexico.

In the mid-1960s, this giant engineering project in water management was seen by leading figures in the U.S. Congress and elsewhere as the next great undertaking to which the United States should commit itself as a nation, comparable in scope and benefits to the NASA space program and the rapid and widespread development of nuclear power. (In fact, the NAWAPA plan was favorably reviewed in the *Bulletin of the Atomic Scientists*.)

*This appendix was excerpted from Lyndon H. LaRouche, Jr.'s 1983 pamphlet,* Won't You Please Give Your Grandchildren a Drink of Water?

## Development is the Name for Peace

In 1964, the Ralph M. Parsons Company, the West Coast-based international engineering firm which had helped to design and build the water management system which turned California into the richest agricultural producing area in the nation, presented a developed plan for NAWAPA to a special subcommittee of the United States Senate chaired by Senator Frank Moss of Utah. As entered in the *Congressional Record*, the original NAWAPA plan called for no less than 369 separate projects.

NAWAPA begins with construction of a series of dams in Alaska and the Canadian Yukon, trapping the water of the various rivers running through this largely undeveloped wilderness area. The drainage area to be tapped is approximately 1.3 million square miles, with a mean annual precipitation of 40 inches.

A large portion of the water thus collected would then be channeled into a man-modified reservoir 500 miles long, 10 miles wide, and 300 feet deep, constructed out of the southern end of the natural gorge known as the Rocky Mountain Trench in the Canadian province of British Columbia. This would be accomplished through a series of connecting tunnels, canals, lakes, dams, and, because the trench itself exists at an elevation of 3,000 feet, even lifts. The network of projects provides plentiful opportunities for hydroelectric power development.

To the east, a thirty-foot deep canal would be cut from the Trench to Lake Superior, to maintain a constant water level and clean out pollution in the entire Great Lakes system from Duluth to Buffalo. Not only would this provide more water for hydroelectric power and agricultural irrigation of the Great Plains region of Canada and the U.S.A., the canal could ultimately be made navigable for lake- and ocean-going vessels from the Great Lakes into the heart of Alberta, and eventually, extended westward into Howe Sound, British Columbia. The dream of a Northwest Passage would at last become a fact, from the Gulf of St. Lawrence to Vancouver.

South from the Trench reservoir, water would be lifted through a giant dump lift to the Sawtooth Reservoir in southwestern Montana, from which point it would flow by gravity through the western part of the system, passing through a tunnel in the Sawtooth Mountain eighty feet in diameter and fifty miles in length, to the western and southern U.S. states.

South of the Rocky Mountain Trench, in central Idaho and southeastern Washington, a series of hydroelectric plants would develop the Clearwater and Clearwater North Fork Rivers and the lower reaches of the Salmon and Snake Rivers. Flow of the Columbia River would be supplemented as needed from other rivers as well as regulated at its direct connection to the Rocky Mountain Trench Reservoir to prevent flooding. NAWAPA aqueducts and reservoirs would dot the slopes of the Rocky Mountains, providing water to the Staked Plains and lower Rio Grande River basin and serving New Mexico, Texas, Colorado, Kansas, Nebraska, Oklahoma and Mexico via existing rivers.

Flows from the Rocky Mountain Trench and Clearwater subsystem would also supply Idaho, Oregon, Utah, Nevada, California, and Arizona in the United States; and Baja California, Chihuahua, and Sonora in Mexico. A diversion aqueduct at Trout Creek, Utah would send high-quality, low-mineral water to southern California and Baja California. Here it would arrest soil damage caused by high-mineral Colorado River irrigation water.

The 1964 Parsons Company study estimated that NAWAPA could assure adequate water supply to the continent for the next 100 years. The conserved water would be sufficient to irrigate 86,300 square miles, equal to a 35-mile-wide strip extending 500 miles into the Canadian agricultural belt, traversing the length of the United States, and extending 200 miles into Mexico for a total length of 2,500 miles. In delivering 20 million acre-feet of water to Mexico, the plan allows that country alone to develop eight times as much new irrigated land as the Aswan High Dam provides Egypt. In this original proposal, Canada would receive 22 million acre-feet of usable fresh water, and the United States 78 million acre-feet.

For political reasons, the NAWAPA proposal was not acted on by Congress when originally presented. But no one has reasonably challenged estimates of the plan's feasibility.

# The Benefits of NAWAPA

For the United States, the benefits of the upgraded NA-WAPA proposal are virtually unlimited. The full-scale project now promises 150 million acre-feet of water per year—a 50 percent increase in the present consumption of 300 million acre-feet yearly. Some 55,000 megawatts per year of surplus electric power would be provided, nearly doubling present U.S. hydroelectric capacity of 70,000 megawatts. Nearly 50 million more acres of irrigable land will become available, almost doubling irrigated acreage west of the Mississippi.

It doesn't end there. Stabilization and control of the Great Lakes is one dramatic example of the decrease in pollution levels attainable by such methods of water management. NA-WAPA would also help to stabilize water levels throughout the West, providing, among its notable benefits, the opportunity to reverse the depletion of the Ogalala Aquifer, the principal water supply for 11 million acres of prime farmland in Texas, Oklahoma, Kansas, New Mexico, and other High Plains states. NAWAPA would provide the mechanism for reversing the current salinity problem of irrigated lands by flooding selected areas to wash out the accumulated salts, and by maintaining a regime of "wasteful" irrigation to prevent such build-ups in the future. Thus ground water supplies would be recharged. In addition, increased facilities for water transport would also prove cost-saving.

NAWAPA would also create substantial numbers of productive jobs, for example, in the construction and steel industries. Unlike the make-work employment projects proposed by some as a depression measure, the NAWAPA project would employ American manpower to actually increase the national wealth.

In addition, NAWAPA would increase the power of the U.S.A. to develop other nations in the Third World, providing new markets for American agriculture and industry.

It is more difficult to give a "dollars and cents" estimate of NAWAPA's annual benefit to American industry, agriculture, workers and consumers; but one respected congressional supporter of the original NAWAPA plan reported it would increase the annual national income from agriculture, livestock, mining, and manufacturing by approximately $30 billion.

The benefits for Mexico and Canada would be of a similar spectacular order. Canada would enjoy 58 million acre-feet of water and 38,000 additional megawatts of hydroelectric power, and the same kind of irrigation, transport, and clean water benefits accruing to the United States. In particular, the Northwest Passage route would be a vital aid in realizing the vast, untapped development potential of that largely wilderness nation.

As for Mexico, a nation whose rapid agricultural and industrial development is essential to advance the living standards of its 60 million citizens and for whom increased food production ranks as a critical national priority, NAWAPA would produce an additional 40 million acre-feet of water a year, at least tripling its irrigable land, and 4,000 additional megawatts of electric power. The Parsons Company's original estimate of the economic benefit to Mexico was an annual $30 billion increase in national income from agricultural, livestock, mining and manufacturing—the same figure as projected for the United States.

# The Costs of NAWAPA

At first glance, the costs of NAWAPA appear immense. The Parsons Company's original 1964 estimate was $80 billion. The upgraded plan was estimated to cost $130 billion in 1979, excluding the costs of environmental studies and other bureaucratic requirements apart from the detailed engineering plans necessary for the project itself. A partial breakdown in 1979 dollars includes $13 billion for construction equipment, $65 billion in construction labor, and 100,000 tons of copper and aluminum, 30 million tons of steel, and 200 million sacks of cement.

Excluded from the $130 billion figure is the cost of needed local distribution systems such as the connection of the Panamint Reservoir to the Los Angeles water supply. Such connections would be required throughout the continent.

To realize the potential of the 50 million acres of new irrigable farmland would require additional capital investment of perhaps $10 billion. (The costs of center-pivot irrigation, for example, are $200 per acre.)

Finally, local transportation systems will need to be upgraded to move the increased produce; better rail transport is especially vital.

Suppose one assumed a $200 billion figure for implementation of the upgraded plan, with a sizable chunk of the local water management, agricultural, and transport expenses included. Averaged out over the ten- to twenty-year lifespan presently projected for completion of the project, that represents a capital expenditure well within the magnitude of the conceivable. If one merely adopts the Parsons Company estimate of $30 billion annual benefit to the national income, it is apparent that NAWAPA would pay for itself many times over by the time our grandchildren have grown up.

Suppose, for the sake of argument, we were to reject the NAWAPA approach as "too ambitious and expensive," decided to "go it alone" without Canada and Mexico, and attempted to develop water resources piecemeal and at a slower pace. We would quickly find that less ambitious is not necessarily cheaper!

Consider the average cost of Bureau of Reclamation water projects from 1975 to 1979. The Bureau spent $700 million on developing an additional 2 million acre-feet of water annually over this period, for an average cost of $350 per acre-foot. Projecting that to the development of 130 million acre-feet required for the western states, produces a total cost of over $45 billion, without any provision for long-distance transfer to the drought areas.

Consider also the average costs of developing additional electric power—at current rates, an additional 40 gigawatts otherwise provided by the western United States NAWAPA would cost $40 billion to replace.

Thus, at the best possible cost estimates, attempting to replace by local projects the U.S. section of NAWAPA alone, excluding the benefits to (and the contributions from) Mexico and Canada, would cost $85 billion, two-thirds as much as the entire project!

In reality, such piecemeal development would be impossible. That only 2 million acre-feet of water per year were developed in the four years preceding 1979, when previous projects were being completed, points to the fallacy of that belief; at that rate our "local," piecemeal approach would take sixty-five years. Only a national commitment to water development can produce benefits on the scale of NAWAPA. Besides, the greatest sources of available water are not in the continental U.S.A., but in Alaska and Canada!

Suppose we decide not to proceed with NAWAPA. That, too, has its costs, and the ultimate cost would be disastrous: The collapse of the agricultural system which has made the U.S.A. the recognized world leader in food production. Take, for example, the problem of the Ogalala Aquifer in the High Plains states, that we mentioned earlier. It is currently estimated that at present rates, this vital source of supply for 11 million acres of farmland will run dry by the year 2020. Increasing sums are being spent on water conservation systems and labor-intensive farming methods to save a few gallons per acre. Not only does this waste capital and human effort, it is leading to an ever-increasing build-up of salts in the soil, salts which will eventually poison the crops and the groundwater if they are not flushed out by "wasting water."

As farmers are forced into dryland farming, the practice being followed in the less arid areas, the effects will be equally disastrous. Dryland farming is necessarily much less productive than irrigated agriculture overall, but it is also *much less predictable*. Therefore, farmers will not be able to invest in crop improvements because they will have much less assurance of a certain level of productivity.

The Great Plains are a major source of food for the world at present, and subjecting this supply to the vagaries of the weather is playing "Russian roulette" with large sections of the world's population. Preliminary studies have shown that the development of irrigation in the once-dry Dust Bowl areas added significantly to the average *natural* rainfall over these areas. If irrigation is now discontinued, it is highly possible that the rainfall will again decline, leaving only dessicated monuments to the greatness of American agriculture.

The only realistic question, then, is, "How shall we proceed with NAWAPA?"

"The management, engineering, and construction of NAWAPA will require the skills of a plethora of organizations," noted N.W. Snyder of the Parsons Company in 1980. He suggested that "some continent-wide agency, representing all three North American countries be formed to finance, build, and operate NAWAPA." Whether this or another approach is taken, one thing remains certain—if the citizens of the United States do not take the lead in discussing and promoting development of NAWAPA, among ourselves and with our neighbors in Mexico and Canada, it will not be built.

# APPENDIX B

# The Development of Africa
### Lyndon H. LaRouche, Jr.

Over the course of the most recent two decades, the most characteristic problem of most so-called developing nations has become the transformation from a "pre-industrial" to a "post-industrial" society, without passing through the "intermediate phase" of becoming an "industrial society." This transformation has been accompanied, in most cases, by a perpetual and worsening dependency upon imports of essential foodstuffs.

By "pre-industrial society," we mean a nation which employs 60 to 90 percent of its total labor force in rural occupations, while most of its urban labor force is either unemployed or employed in unskilled forms of labor-intensive occupations. A "post-industrial society" is one in which more than 60 percent of the urban labor force is unemployed or employed in overhead expense categories. Among developing nations, the

*This appendix is excerpted from* The Independent Democrats' 1984 Platform, *the presidential campaign platform of Lyndon LaRouche and Billy Davis in 1984.*

transformation directly from a "pre-industrial" to "post-industrial" state of affairs takes the form of a significant migration of rural populations to urban centers, to the effect that the increase of the urban labor force in this way increases chiefly the populations of both the unemployed poor and persons employed in administration, reselling, and labor-intensive services, with relatively little expansion of employment in capital-intensive industry.

The effects of such shifts are worsened by the fact that rural production remains more or less "traditional," labor-intensive, and of correspondingly low productivity, in mode.

If such nations are to survive, and the preconditions for durable forms of democratic institutions to be established, this pattern must be broken. It is indispensable that two conditions be met: First, that introduction of modern technology produce a continuing increase in per hectare and per man-year yields in agriculture; and, second, that more than 50 percent of the urban labor force be employed as productive operatives in combined modes of infrastructure development, mining, manufacturing, and construction—which rely increasingly on capital-intensive advances in technology.

There is currently in progress in Egypt a successful program of development which illustrates the proper objectives of U.S. economic foreign policy.

The Egyptian government is implementing a program to double the arable land under cultivation in that nation. By aid of large-scale water-management systems, Egypt is constructing new agro-urban complexes. Large new tracts of cultivation are being developed around newly created cities. The cities, designed to house hundreds of thousands, are the centers of the new complexes, like the hub of a wheel. Around this hub are the industries which employ the city's labor force. Surrounding this urban-industrial complex is the farming area. The farming area employs the American pivot method of irrigation for crop cultivation. In the areas beyond the circles irrigated by the pivot-systems, fruit trees are raised by aid of drip irrigation methods.

The urban centers provide housing that is modest in number and size of rooms, but otherwise modern. Excellent city planning places shopping facilities and light industry efficiently for city-dwellers. Outside the city, good management has often, already transformed the yellow sand of the desert into brown soil. In those sites, within a few years, where there is brown land today, there will be rich black soil.

According to official reports, more than 90 percent of the investment in such projects is internally financed by Egypt.

This example implies the principles of successful economic development of a "developing" nation. It illustrates the relationship of development of basic economic infrastructure and investments in agricultural and industrial capital. It also illustrates the point that, in most developing nations, there are sufficient numbers of trained professionals and other national resources within the nation to contribute the greater part of a project, and that technical assistance and capital supplied by foreigners functions as the critical added margin needed to enable the job to be done.

# How the U.S. Economy Benefits From Such Policies

There is both an immediate and a longer-term benefit to the U.S.A.'s internal economy from exports of capital goods and engineering services as development assistance to "developing" nations.

In the first phase, the immediate benefit is increased production and sales by our own capital-goods-producing industries. By accelerating the turnover of capital investment in these industries, we accelerate the rate at which those industries are able to incorporate technological advances into the design of the capital goods they produce. This means faster rates of improvement in the quality of capital goods those industries produce for our own domestic firms. This improvement, in turn, raises the average level of our national productivity at home.

*The most important source of profit to the United States from capital goods exported is not the profit on the sale. The most important source of profit is the effect of increased capital turnover in the exporting industries.* They become more efficient, through increased volumes of production and sales; more rapid technological advances in those industries increase the productivities in all U.S. industries purchasing capital goods from those firms.

The increased margins of profit this contributes to the domestic economy of the United States more than pays the cost of financing the exports.

Second, not only are we paid back for the credit we extend to importing nations; those nations' cost of producing the goods we import from them is lowered. *As they increase their average productivity of labor, it costs those nations fewer average man-hours to produce each unit of the products we import from them. We benefit from the lowered prices of those imports.*

At the same time, as developing nations increase both their gross national output, and the output (and income), per capita, of their populations, those nations' purchasing power is increased. Their ability, and desire, to purchase increased volumes of capital-goods imports is increased. Our foreign markets grow in this way.

Everyone participates in the trading benefits equitably.

## The Development of Africa

*Approximately 120 million persons in black Africa are presently faced with death from famine and epidemic. An estimated 60,000 black Africans are dying each day of causes attributable to food shortages.* Although the extremes of poverty among these new nations is a relic of European colonialism and the earlier direct and chain-reaction effects of 2,000 years of the slave-trade, the present peril of black Africa is caused by the coincidence of prevailing monetary and international banking policies with the policies of influential Anglo-Saxon eugenicists and other Malthusians who welcome a solution to the alleged "over-population" of black Africa.

*In fact, Africa is the world's most under-populated continent.*

212

# Development of Africa

In the north of Africa, the present dictatorship of Libya is the chief instrument within the continent threatening the stability and continued existence of the nations of Morocco, Algeria, Tunisia, Egypt, Sudan, and many among the states of the Saharan and sub-Saharan region. The susceptibility of increasing portions of the populations of these nations, to insurgencies mediated through the Libyan dictator, is fostered by the spread of despair, of cultural pessimism. This despair is nourished by persistence of conditions which have existed and generally worsened since the middle of the 1960s. Newly liberated former colonies, which once hoped to participate in technological progress, are being driven into worsening conditions instead. This feeds despair, and the cultural pessimism of despair fosters susceptibility to the influence of dark forces of murderous irrationalism.

Needed emergency actions to halt the genocide, and medium- to long-term measures to develop the entirety of this continent, are complementary.

Immediately, to stop the extinction of tens of millions of Africans by famine and famine-fostered epidemics, we must bring emergency food relief. The additional problem is, that even if we bring such food relief to the ports of Africa, rarely do we find the means to transport this food into the locales it is desperately wanted.

Heretofore, emergency food relief to some stricken parts of black Africa has tended to create food-distribution sites in such a manner that hungry Africans were given no alternative but to trek by foot long distances to reach those sites. Such treks often became "death marches." The survivors assembled around the sites in encampments remind us somewhat of the refugee camps in Europe at the end of the last war. Food relief so conducted wrecked the fragile existing infrastructure and social organization of the affected portions of the populations.

The objectives of food relief must be:

1) To bring the food relief to the locales in which the needing population lives and works;

2) To supplement food relief with measures which aid in

restoring the affected villages and their inhabitants to dignified self-subsistence.

The required methods are those with which some of us are familiar from U.S. engineering feats during World War II. We must build emergency ports, emergency rail lines, and emergency roadways, as U.S. engineers did during that war. We must use the logistical methods of general warfare for the great emergency works of peace.

The proper locations and routes of such emergency works of transportation will be the same locations and routes which must be developed and made functional as part of the long-term development of the same regions. The proper location for the emergency port will be the same place that nation needs a port in any case. The emergency rail line, the emergency roadway, will lie along the same route a permanent railway or road is required. The emergency efforts to develop a water-supply will be the same required to prevent the catastrophe from recurring. The military and other forces of African nations drawn into the effort will be practicing the methods and procedures the nation requires of them in further development and maintenance of basic economic infrastructure.

We Americans used to know how to accomplish what seemed miracles along those lines. If those skills have become rusty, our military and civilian task forces suited to such work reduced in scale, it is past time we restored ourselves to excellence in such matters. Let us put our Pentagon, our Army Corps of Engineers to work. Mobilize from ourselves, and our friends in this hemisphere, in Europe, and elsewhere, the ships, the fleets of helicopters, the trucks, to save lives now where desperately hungry people live. Accompany this first phase with the beginning of a second: show the miracles our ingenuity and stubborn wills can accomplish in creating emergency ports, cutting emergency roads, laying rail lines, moving water-supplies.

In the process, help these stricken nations to help themselves. Train their nationals to assist and continue the work, and phase ourselves out as those nationals take over the work of completing the task. Help get the crops planted, and set the needy back on their own feet.

For the longer term, Africa's most fundamental needs are the most basic of economic infrastructure and increased production of food: water-management, transportation, energy, and agriculture.

The heart of the solution for the urgent needs of the continent as a whole is a rail system cutting across the Sahel, preferably from Dakar in the west, to Djibouti in the east, and a redistribution of water, from the surplus water of the rain-forest region of north-central Africa into the central Sahel. The water-supplies must come from management of the Zairean and Victoria region and southern Sudan. The first line distributes water to the region around now-dying Lake Chad; the second is a program of cooperative water-management of the Nile system. The east-west rail line (actually projected as early as the 1870s!) intersects the existing railway systems of Nigeria and Egypt-Sudan. The rail network must then be extended by north-south intersecting trunk lines: north-south from Algeria, and south into Tanzania. Without these two sets of measures of water-management and railway development, the rational economic development of the continent as a whole is impossible.

The east-west rail line, across the Sahel, serves as the indispensable logistical base-line for deploying to reverse the present spread of the Saharan desert into the Sahel. By joining this with north-south links, this rail system plays a vital part in fostering initially modest but crucial trade among the nations of both Arab and black Africa. The combined effects of railway development and water-management are optimal exploitation of combined railways and navigable waterways, creating the beginnings of a functioning internal transport system. This development, enriched by development of the beginnings of a continent-wide system of energy production and distribution, provides the logistical basis to begin the improvement of agriculture throughout much of Africa, in depth. Politically, it aids African nations, by strengthening the benefits each local sector of the population enjoys through aid of the government's central authority.

## Development is the Name for Peace

On principle, there is little really new in such policies. These are proven policies which Europe and the Americas, and others before them, developed over thousands of years of cumulative experience.